Praise for

USE YOUR BRAIN TO CHANGE YOUR AGE

and Dr. Daniel G. Amen

"I have been a longtime advocate of the pioneering research of Daniel Amen. I believe this book may be his best ever as it distills much of his previous work coupled with some of the newest research into a comprehensive, but clearly stated lifestyle program. The science behind the program is complex, but its execution is easy to follow if you truly want to separate biological age from chronological age."

—Barry Sears, Ph.D., author of *The Zone*

"*Use Your Brain to Change Your Age* is inspiring and practical. Its case studies make this book powerfully impactful, both for adults and young adults in schools. I can't think of a single reader who won't find himself or herself represented in the stories of women and men who have used Dr. Amen's wisdom and research to live healthy lives. This is a must-read for anyone interested in the human brain and the body that shelters it."

—Michael Gurian, author of *Leadership and the Sexes* and *Boys and Girls Learn Differently*

"Thorough, practical, and inspiring, this is an essential guide to reclaiming your brain—and your life! It's never too late to start . . . or too early. I recommend it to anyone with a brain."

—Hyla Cass, M.D., author of *8 Weeks to Vibrant Health*

"Your brain health is intimately tied to your physical health, and in this revolutionary and life-transforming book you will learn to help the organ that the majority of the medical, psychiatric, and psychological health-care professionals neglect—the brain! Whether you are a first-time reader of Dr. Amen or someone like me who has read all of his books, you will learn new insights and will feel motivated to make the changes in your life to become a younger, better-looking you. If you change your brain you will not only change your life, you will improve the quality and length of your life!"

—Earl R. Henslin, Psy.D., author of *This Is Your Brain on Joy*

"I couldn't stop reading Dr. Amen's new book until I was finished. As a holistic neurosurgeon, I found it to be the best user's guide to optimal brain function I have ever read. His incredible storytelling skill combined with practical information will change your life."

—Joseph C. Maroon, M.D., professor and vice chairman of the Department of Neurosurgery, University of Pittsburgh Medical Center, and team neurosurgeon, Pittsburgh Steelers

---

"This book is so engaging that I almost burned my dinner while reading it. . . . The overarching message is hope and motivation. As Dr. Amen proposes, the fountain of youth is between your ears."

—Ingrid Kohlstadt, M.D., Ph.D., *Townsend Letter*

"Obesity, depression, and Alzheimer's disease are current epidemics that are predicted to get worse. If you want to avoid them and improve your physical and mental health, read Dr. Amen's books."

—Stephen R. Covey, author of *The 7 Habits of Highly Effective People* and *The Leader in Me*

"*Use Your Brain to Change Your Age* provides outstanding and practical advice based on what we know about how the brain and our body work. The information in this book is tremendously important for anyone wanting to stay young and keep their body and brain as young and healthy as possible. Important for everyone whether you are older and want to grow younger or younger and want to get a head start on living a long and vigorous life."

—Andrew Newberg, M.D., coauthor of *How God Changes Your Brain* and director of the Myrna Brind Center of Integrative Medicine, Thomas Jefferson University

---

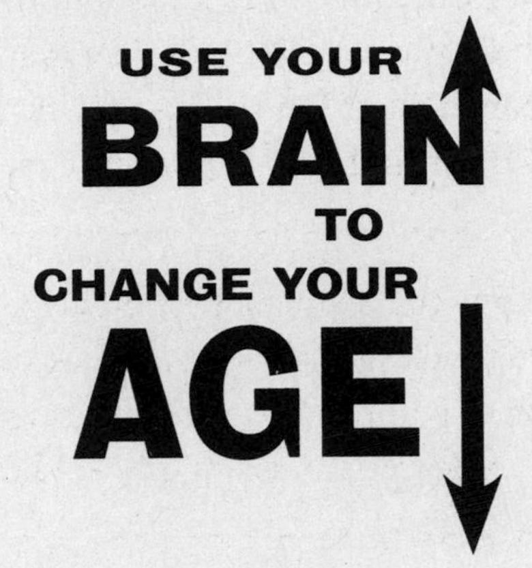
USE YOUR
BRAIN
TO
CHANGE YOUR
AGE

ALSO BY DR. DANIEL G. AMEN

THE AMEN SOLUTION (Crown Archetype, 2011, *New York Times* Bestseller)

END EMOTIONAL OVEREATING NOW (written with Larry Momaya, M.D., MindWorks Press, 2011)

UNCHAIN YOUR BRAIN (MindWorks Press, 2010)

WIRED FOR SUCCESS (MindWorks Press, 2010)

CHANGE YOUR BRAIN, CHANGE YOUR BODY (Harmony Books, 2010, *New York Times* Bestseller)

MAGNIFICENT MIND AT ANY AGE (Harmony Books, 2009, *New York Times* Bestseller)

THE BRAIN IN LOVE (Three Rivers Press, 2007)

MAKING A GOOD BRAIN GREAT (Harmony Books, 2005, Amazon Book of the Year)

PREVENTING ALZHEIMER'S (written with neurologist William R. Shankle, Putnam, 2004)

HEALING ANXIETY AND DEPRESSION (written with Lisa Routh, M.D., Putnam, 2003)

NEW SKILLS FOR FRAZZLED PATIENTS (MindWorks Press, 2003)

HEALING THE HARDWARE OF THE SOUL (Free Press, 2002)

IMAGES OF HUMAN BEHAVIOR: A BRAIN SPECT ATLAS (MindWorks Press, 2003)

HEALING ADD (Putnam, 2001)

HOW TO GET OUT OF YOUR OWN WAY (MindWorks Press, 2000)

CHANGE YOUR BRAIN, CHANGE YOUR LIFE (Three Rivers Press, 1999, *New York Times* Bestseller)

ADD IN INTIMATE RELATIONSHIPS (MindWorks Press, 1997)

WOULD YOU GIVE 2 MINUTES A DAY FOR A LIFETIME OF LOVE (St. Martin's Press, 1996)

A CHILD'S GUIDE TO ADD (MindWorks Press, 1996)

A TEENAGER'S GUIDE TO ADD (written with Antony Amen and Sharon Johnson, MindWorks Press, 1995)

MINDCOACH: TEACHING KIDS TO THINK POSITIVE AND FEEL GOOD (MindWorks Press, 1994)

THE MOST IMPORTANT THING I LEARNED IN LIFE I LEARNED FROM A PENGUIN (MindWorks Press, 1994)

TEN STEPS TO BUILDING VALUES WITHIN CHILDREN (MindWorks Press, 1994)

THE SECRETS OF SUCCESSFUL CHILDREN (MindWorks Press, 1994)

HEALING THE CHAOS WITHIN (MindWorks Press, 1993)

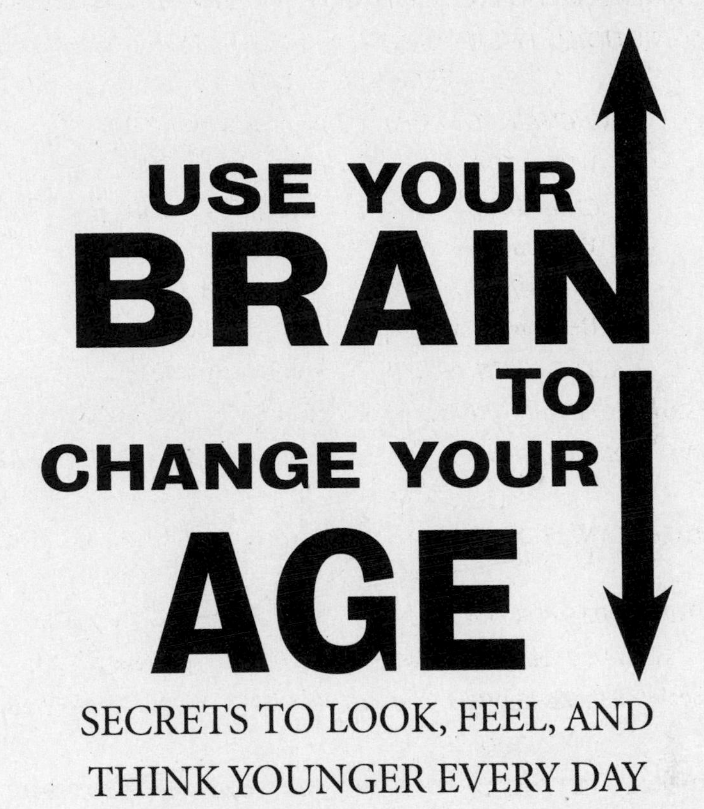

SECRETS TO LOOK, FEEL, AND THINK YOUNGER EVERY DAY

DANIEL G. AMEN, M.D.

CROWN ARCHETYPE
NEW YORK

MEDICAL DISCLAIMER

The information presented in this book is the result of years of practical experience and clinical research by the author. The information in this book, by necessity, is of a general nature and not a substitute for an evaluation or treatment by a competent medical specialist. If you believe you are in need of medical intervention please see a medical practitioner as soon as possible. The stories in this book are true. The names and circumstances of the stories have been changed to protect the anonymity of patients.

Published in the United States by Crown Archetype, an imprint of the
Crown Publishing Group, a division of Random House, Inc., New York.
www.crownpublishing.com

Crown Archetype with colophon is a trademark of Random House, Inc.

Library of Congress Cataloging-in-Publication Data is available upon request.

ISBN 978-0-307-88854-9
eISBN 978-0-307-88856-3

Printed in the United States of America

Book design by Lenny Henderson
Jacket design by Michael Nagin
Author photograph: Blake Little

10 9 8 7 6 5

First Edition

To Tana, my reason for living a long, healthy life!

CONTENTS

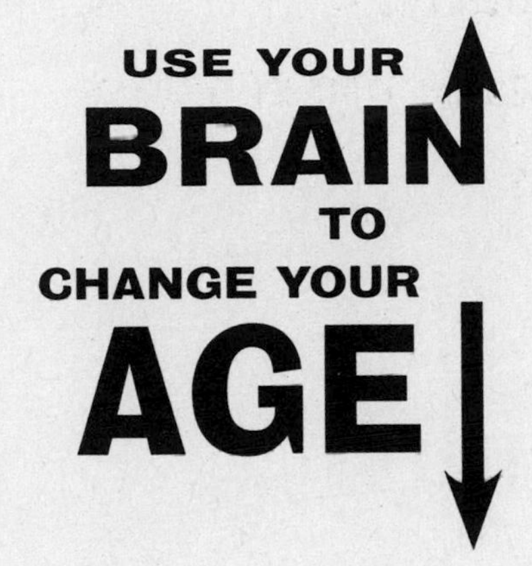
USE YOUR
BRAIN
TO
CHANGE YOUR
AGE

INTRODUCTION

THE FOUNTAIN OF YOUTH IS BETWEEN YOUR EARS

SEVEN PRINCIPLES THAT WILL CHANGE EVERYTHING IN YOUR LIFE

You are only young once,
but you can stay immature indefinitely.
—Ogden Nash

I was recently on a plane from San Francisco to Honolulu, where I was going to participate in a very important debate at the annual meeting of the American Psychiatric Association. In the seat next to me was an elderly woman, Mary, who recognized me from my public television shows. As I was opening my computer to start obsessing over the debate, Mary leaned toward me and asked, "Is it ever too late?"

"Too late for what?" I said, trying to get my mind focused on the task tomorrow.

"I am seventy-six years old," she whispered. "Is it ever too late for me to have a better brain?"

"Only if you just plan on living until you are seventy-seven," I said with a smile, now looking into her pretty green eyes. "If you want to live until you're ninety, *now* would be a good time to start!"

She chuckled. I relaxed. People like Mary have always fueled the passion I have for my work.

"*The fountain of youth is between your ears,*" I continued. "It is your brain that makes the decisions that keep you healthy, happy, and on track to live a long time, and it is your brain that makes the bad decisions that ruin your health and send you off to the undertaker early. If you want to live a long, happy life, the first place to start is by having a better brain."

Mary told me she loved my television shows because they were so practical and she had already made many changes in her life. She also told me about her son who was a problem drinker but who stopped imbibing after watching my programs. He saw he wanted no part of the damage he saw on the SPECT scans that alcohol can do to the brain.

At the Amen Clinics we use a sophisticated brain imaging study called SPECT to help us understand and treat our patients. SPECT stands for *single photon emission computed tomography*, a nuclear medicine study that looks at blood flow and activity patterns. It looks at how the brain works. It is different from computerized axial tomography (CAT) or magnetic resonance imaging (MRI) scans, which are anatomical scans that show what the brain physically looks like. SPECT shows how the brain functions. Over the last twenty-one years, Amen Clinics has built the world's largest database of brain SPECT, now totaling over seventy thousand scans on patients from ninety different countries.

SPECT basically shows three things:

1. Areas of the brain that work well
2. Areas of the brain that are low in activity
3. Areas of the brain that are high in activity

A healthy scan shows full, even, symmetrical activity in the brain.

From looking at all of these SPECT scans it is very clear to me that *you can either accelerate the aging process and make your brain look and feel older than your chronological age, or you can decelerate it and have a brain that looks and feels much younger than your age.*

Even though getting older is not optional,
having a brain that looks and feels old is!

Here are three SPECT scans of sixty-year-old brains. One is healthy, one has Alzheimer's disease, and one is from a person who is overweight and has sleep apnea.

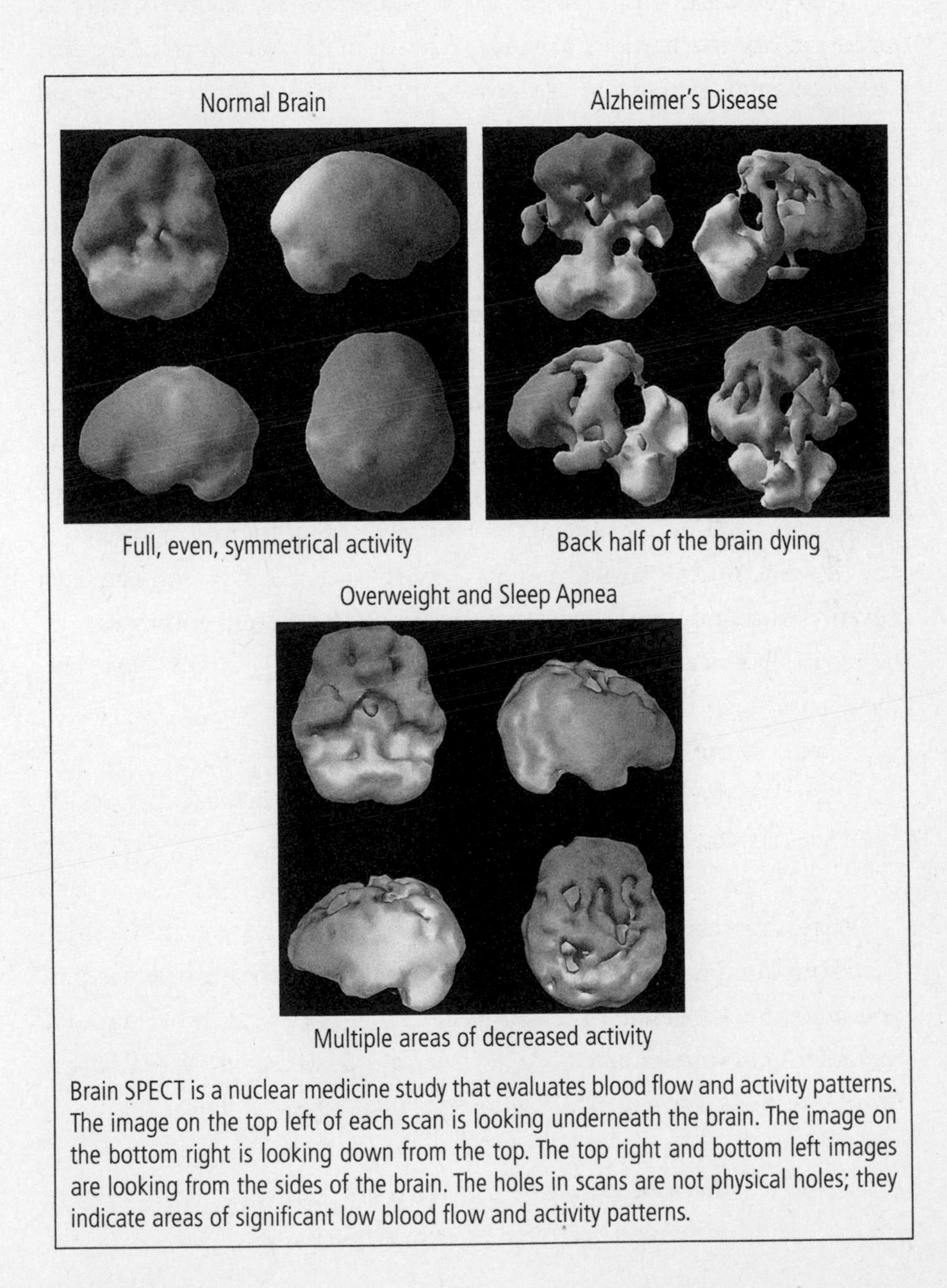

Brain SPECT is a nuclear medicine study that evaluates blood flow and activity patterns. The image on the top left of each scan is looking underneath the brain. The image on the bottom right is looking down from the top. The top right and bottom left images are looking from the sides of the brain. The holes in scans are not physical holes; they indicate areas of significant low blood flow and activity patterns.

Of these three brains, which do you want? Who will live the longest and have the youngest and most effective brain? The choice should be obvious.

It is very clear from our imaging work that as we age brain activity decreases across the whole surface of the brain. If we are not thoughtful in how we live, the myriad bad decisions a person makes over a lifetime influence aging in a negative way. Lousy diets, chronic stress, health problems, a lack of sleep, too much alcohol, illegal drugs, high-risk behavior, being around environmental toxins, and many other factors contribute to our brain's early demise. Unfortunately, most people just accept a decline in cognitive functioning as normal aging.

I recently recorded an interview with a high-level business executive, Todd. He told me his memory was terrible at age fifty-three. "I am sure it is just my age. I am just getting older," he said. "I often have no idea where I put my keys and sometimes find them in the refrigerator, next to the eggs."

"It is definitely not normal," I replied. "I am fifty-seven and my memory is as good as it has ever been. It is one of the little lies people tell themselves to justify their memory problems and bad habits. The denial prevents them from getting the help they need. Tell me about your diet and exercise."

When Todd heard me mention exercise, he perked up. "I exercise five times a week. I run long distances and am in great shape."

Something wasn't making sense to me. "And your diet," I persisted.

He looked down. "It's not so great. Every morning I have a Diet Coke and Pop-Tarts in the car on the way to work. The rest of the day doesn't get much better."

Putting toxic fuel in a car will definitely decrease its performance. Putting toxic fuel in your body will definitely hurt your brain, no matter how much exercise you do.

"If you had a million-dollar racehorse," I asked, "would you ever give it junk food?"

"Of course not," he said.

"You are so much more valuable than a racehorse. It is time to treat yourself with a little love and respect," I encouraged.

Three months later, Todd told me his memory had significantly improved. He also said I haunt him at every meal. I am hoping to do the same for you.

Bill, age eighty-five, had a typical brain scan for someone his age. He was a retired business executive and complained of being tired. He struggled with his memory and was on four medications for hypertension, cholesterol issues, and chest pain. "I hate getting older," he said. (The language we use around getting older, as we will see, is very important.) His brain SPECT, not surprisingly, depicted an old brain, with overall low activity.

Bill's Scan, Typical for an 85-Year-Old

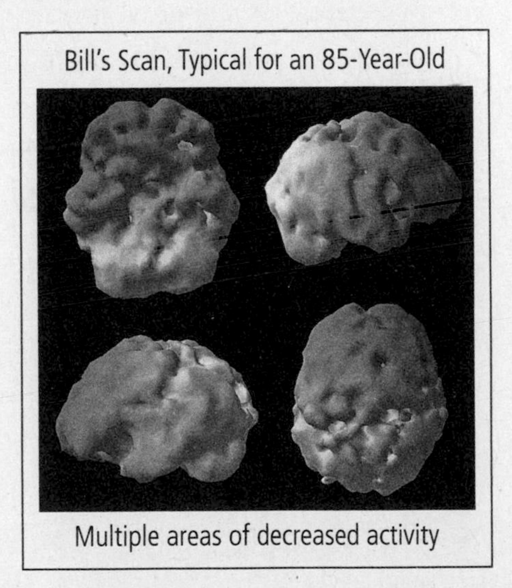

Multiple areas of decreased activity

Through our brain imaging work, we have also discovered a group of older men and women who have stunningly beautiful-looking and functioning brains, and they all have lives that reflect their healthy brain function. Their lives are much more vibrant, energetic, and thoughtful than those with poorer-functioning brains. In addition, they all maintain many of the brain healthy habits that you will learn about in this book. Interestingly, you never hear them saying they hate getting older. They appreciate their experiences and relationships, and they look forward, not backward.

Dr. Doris Rapp is a great example. She is an eighty-two-year-old pioneering physician who has dedicated her life to helping others. She has been called the "Mother of Environmental Medicine and Allergies." She maintains a healthy weight, exercises vigorously, eats a nutritious diet, and has spent her life pursuing learning opportunities. She has a sharp mind and many friends, and she still consults with patients and professionals. I often call her when I am stumped with a patient, especially if I suspect a problem with environmental toxins or food allergies. Both her SPECT scan and her life reflect her healthy-looking brain. If you make brain smart decisions, you can definitely slow and, as we will see, in many cases reverse the aging process.

Doris, 82

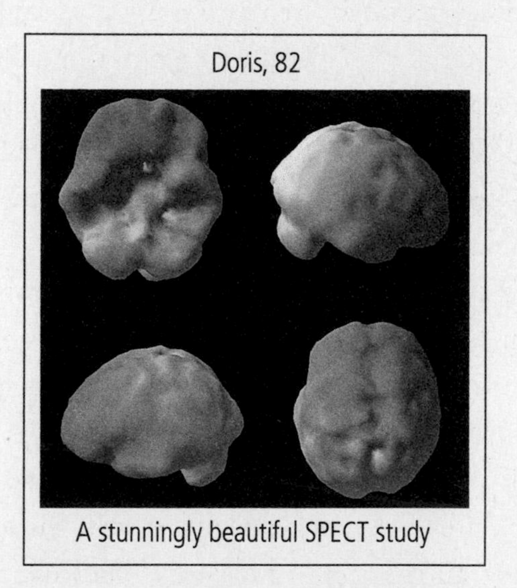

A stunningly beautiful SPECT study

In this book I am going to show you how to have a younger-looking brain and body. It is based on the lessons that we have learned at the Amen Clinics. If you follow the steps in the program you will dramatically increase your chances of living longer; looking younger; decreasing your risk of dementia; and boosting your memory, mood, attention, and energy. I know this is a big promise, but I have seen that when people engage in this process, everything in their life changes in a positive way. You just have to spend some time to "dial in" the program and make it part of your everyday life, but once you do, the benefits will be lifelong.

CARLOS

When we first saw Carlos, he was forty-eight years old and filled with worries, negative thinking, depression, and anger, and he had trouble focusing. He had undiagnosed dyslexia as a child and struggled with heavy drinking in the past. His health habits were terrible, he weighed 266 pounds, and his brain was in trouble, all of which were not helping his emotional issues. Below is Carlos's initial brain SPECT scan.

Carlos completely bought into the program we laid out for him. He is an analytical man, so the logic of the program made sense to him. It is the same program I am going to give you in this book. After ten weeks he lost 24 pounds, and after thirty weeks he was down 50 pounds. More important, his mood, energy, and memory were better as well. And he looked and felt ten years younger.

By learning and doing the techniques in the program, he no longer overate to medicate his sadness and irritability. And by eating brain healthy foods at frequent enough intervals, he no longer had the energy crashes that made him so vulnerable to stress.

He looks like a different person on the outside, but we saw the same dramatic differences on the inside as well. His follow-up SPECT scan showed overall increased activity. By following this program, he changed his brain and in the process changed his life!

This is the part I love most about Carlos's story. After seeing her husband's success, his wife, who was not overweight, started our program to learn how to create a brain healthy family and ended up losing 10 pounds herself. Then his fourteen-year-old daughter embarked

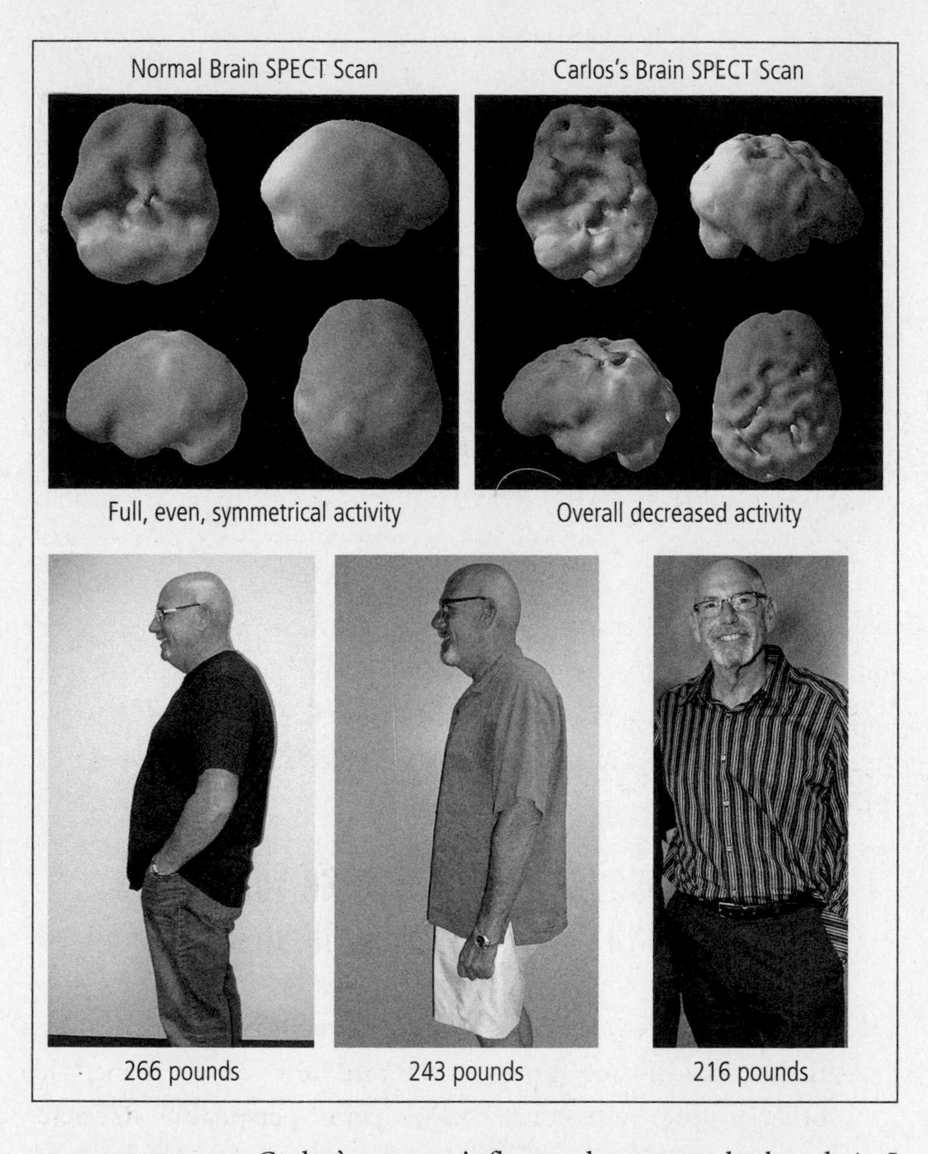

Full, even, symmetrical activity

Overall decreased activity

266 pounds

243 pounds

216 pounds

on our program. Carlos's success influenced everyone he loved. As I was writing this book I saw Carlos in our waiting room. Two years after I first met him he still looked fabulous. I asked him how he kept it going.

"It's not hard," he said. "I have the program dialed in."

You can do this too. None of what I will ask you to do is hard. It just takes consistent effort.

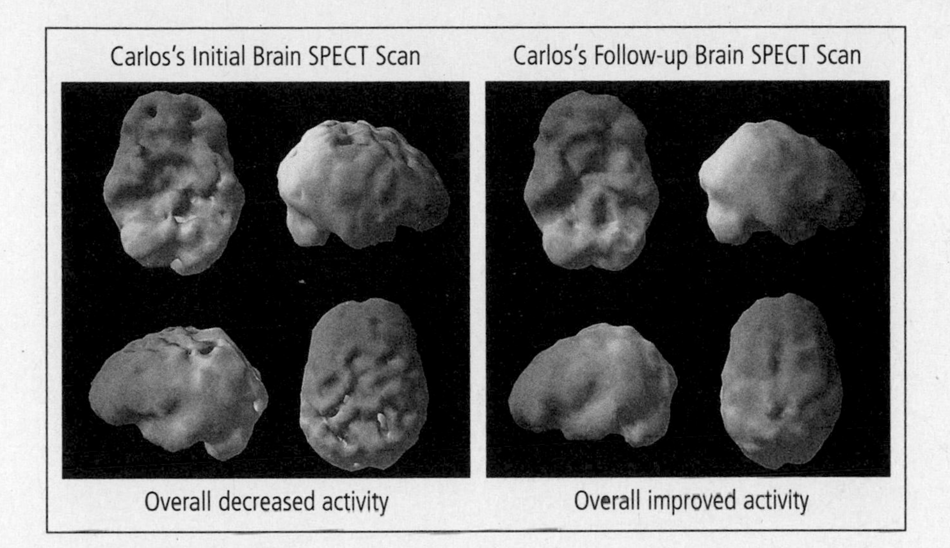

The brain imaging work that we do at the Amen Clinics has guided our practice for more than two decades. Over the years I have crystallized our work into seven very simple principles. These principles guide all of the work we do, and they provide the foundation for this program.

SEVEN PRINCIPLES FOR CHANGING YOUR BRAIN AND IMPROVING EVERYTHING IN YOUR LIFE

1. **Your brain is involved in everything you do, including how you think, how you feel, how you act, and how you get along with other people.** Your brain is the organ of personality, character, intelligence, and every decision you make. And, as we will see, *it is the quality of your decisions that help you live a long time or kill you early.*

    - It's your brain that tells you to stop eating when you've had enough . . . or lets you dive into that second bowl of ice cream so that you end up feeling bloated, groggy, and depressed.

- It's your brain that reminds you to drive carefully . . . or urges you to race down the highway at speeds that can cause tickets, crashes, or fatalities.
- It's your brain that keeps you focused, motivated, and successful . . . or holds you back with attention or anxiety issues.
- It's your brain that keeps you calm, happy, and loving . . . or disrupts your moods and creates conflicts in your relationship.
- It's your brain that works as the "control center" for your entire physical system.
- And it's your brain that deserves credit, attention, and care for everything it does to run your body, your mind, and your life!

2. **When your brain works right, you work right; when your brain is troubled, you are much more likely to have trouble in your life.**

With a healthy brain you are

- Happier
- Healthier
- Wealthier
- Wiser
- More effective
- Able to make better decisions, which helps you live longer

When the brain is not healthy, for whatever reason, you are

- Sadder
- Sicker
- Poorer
- Less wise

- Less effective
- More likely to make bad decisions

3. **The brain is the most complicated organ in the universe and thus the most vulnerable to damage and aging.** There is nothing as complicated as the human brain. Nothing. It is estimated that the brain has a hundred billion cells. Each brain cell is connected to other brain cells by thousands of individual connections between cells, which means that you have more connections in your brain than there are stars in the universe! A piece of brain tissue the size of a grain of sand contains a hundred thousand neurons and a billion connections all communicating with one another.

   The brain is 80 percent water. Hydration is critical to the brain's health. The solid weight of the brain is 60 percent fat, so any abnormalities in the body's fat content can wreak havoc on the brain. Even though your brain consists of only 2 percent of the body's weight, it uses 20–30 percent of the calories you consume. Of the breakfast you had this morning or the dinner you had tonight, approximately a quarter of it went to feed your brain. Your brain also consumes 20 percent of the oxygen and blood flow in the body and it never rests (even during deep sleep). Because of its high metabolic rate, it produces a high level of free radicals, which can damage the brain, if the antioxidant capacity of the brain is low. *Your brain is the most energy hungry and expensive real estate in your body.* And, for good reason, it is the command and control center that runs your life.

4. **Your brain is very soft, about the consistency of soft butter, tofu, or custard, and it is housed in a hard skull, which has multiple sharp, bony ridges.** Brain injuries matter. They can ruin people's whole lives. Very few people understand just how much brain injuries matter, because brain imaging is only just now starting to

gain wider usage. In a 2008 front-page article in the *Wall Street Journal,* writer Thomas Burton quoted researchers who found that undiagnosed brain injuries are a major cause of the following:

- Homelessness
- Psychiatric illness
- Depression and anxiety attacks
- Alcoholism and drug abuse
- Suicide
- Learning problems

If you want to stay healthy, protecting your brain should be one of the first things you do.

5. **Our brain imaging work has clearly shown that there are many things you do to accelerate the aging process, making your brain look and feel older. And there are many things you can do to decelerate the aging process, helping you and your brain look and feel younger.** This one fact, based on tens of thousands of scans we have seen at the Amen Clinics, is the major reason I am writing this book. People need to know that every day through their behavior they are either helping or hurting the health of their brain.

Most people have a lot more say in their long-term health than they give themselves credit for. Certainly, having the right combination of genes can be helpful, but most genes get turned on or turned off based on your behavior. According to recent research, only about 30 percent of longevity is determined by your genetics. The other 70 percent is up to you. Your habits determine how old your brain is and, subsequently, how long you will live and how well you will live.

The bottom line of brain health can be summed up in four words:

AVOID BAD • DO GOOD

Of course, the details underlying these four words require a little more detail.

Here is a list of problems and behaviors that accelerate the brain's aging process and can take years off your life. If you want to live a long time with your brain intact, avoid these as much as possible:

- Inconsistent, thoughtless behavior and decisions that negatively affect your health
- Unhealthy friends or a lack of a positive support system. The people you spend time with matter. People are contagious, and if you spend time with unhealthy people, you are much more likely to pick up their brain-damaging ways. This doesn't mean you get rid of all your friends and family if they have unhealthy habits, but limit the amount of time you spend with them and get a new, healthier group if you want to live long.
- Brain injuries
- Toxins
    - Drugs, illegal and many legal drugs, such as benzodiazepines and pain killers
    - Alcohol, more than a few glasses a week
    - Smoking
    - Excessive caffeine, more than 300 mg (milligrams) a day (three normal-size cups of coffee)
    - Environmental toxins, such as pesticides, organic solvents, phthalates, and mold

- o Cancer chemotherapy or chemo brains. As chemotherapy kills cancer cells, it can also be toxic to normal cells as well. If you have had or need to have chemotherapy (please discuss this with your health-care provider), make sure to embrace a brain healthy life.
- Inflammation. Inflammation that becomes chronic is now considered to be a major cause of many diseases of aging, including cancer, diabetes, heart disease, and Alzheimer's. Inflammation is promoted by free radicals formation, low levels of vitamin D or omega-3s, high levels of omega-6s, high-meat and/or high-sugar diets, diabetes, long-term infections, gum disease, and stress.
- Exposure to free radicals, or molecules that can cause damage in the body. Not unlike the way rust attacks a car, free radicals attack our cells, damage our DNA, and accelerate aging. Things to avoid: cigarettes, trans fats, excessive sun exposure, charred meats, pesticides, excessive exercise, overactive thyroid gland, and inflammation. Although fruits and vegetables are great sources of antioxidants, which fight free radicals, avoid buying the "dirty dozen" (produce with the highest chemical residue: peaches, apples, blueberries, bell peppers, celery, nectarines, strawberries, cherries, imported grapes, spinach, kale, and potatoes), and buy their organic counterparts instead.
- Damaged DNA and telomeres. At the end of each long strand of DNA is a cap called a telomere, which is much like the plastic seal on the end of a shoelace. The DNA cap's purpose is to keep DNA from unraveling. Each time a cell divides, a little piece of the telomere is eroded. After about sixty divisions, it is completely gone, allowing the DNA to unravel. Inflammation, free radicals, vitamin deficiencies, and a lack of omega-3 fatty acids can all chip away at the telomere, shortening the

cell's life span. Geneticist Richard Cawthon and colleagues at the University of Utah found that shorter telomeres are associated with shorter lives. Among people older than sixty, those with shorter telomeres were three times more likely to die from heart disease and eight times more likely to die from infectious disease.

- Medical problems
  - Gum disease
  - Heart disease
  - Diabetes
  - Hypertension
  - Gum disease
  - Intestinal or stomach problems
  - High or low testosterone hormone levels
  - High or low thyroid hormone levels
  - Low omega-3 levels
  - Low vitamin D levels
  - High iron levels, which increase oxidative stress
  - Allergies
  - Chronic insomnia or sleep apnea
- Unhealthy weight gain or obesity. As your weight goes up, the size of the brain goes down. (That should scare the fat off anyone!)
- Standard American diet
  - Sugar. When too much sugar mixes with proteins and fats, it forms molecules called advanced glycation end products, which promote aging. The American Heart Association now recommends that women consume no more than 100 calories per day of added sugars and that men consume no more than 150 calories per day.
  - Trans fats
  - Excessive calories

- Lack of exercise, endurance, and strength
- Lack of new learning
- Mental health issues and/or chronic stress
  - o Depression
  - o Negative thinking patterns
  - o Excessively high or excessively low levels of anxiety
  - o Excitement seeking or impulsive behavior
  - o Negative messaging around aging
- Lack of brain rehabilitation strategies when needed
- Lack of proper supplementation or the indiscriminate use of supplements
- Lack of meaning and purpose in your life
- Lack of knowledge on how your brain is functioning

BUT HOW CAN I HAVE ANY FUN?

We have a high school course called Making a Good Brain Great that is in forty-two states and seven countries, which teaches young teenagers how to take care of their brains. After we teach the part of the course on things to avoid in order to get and keep a great brain, invariably an "attitudinal" teenage boy will blurt out in class, "So, how can I have any fun?" We then do an exercise with the class called "Who has more fun, the person with the good brain or the person with the bad brain?"

Who gets the date with the pretty girl and gets to keep her because he doesn't act like an idiot? The guy with the good brain or the guy with the bad brain? The guy with the good brain!

Who has more freedom because he has more consistent behavior and his parents trust him more? The teen with the good brain or the teen with the bad brain? The teen with the good brain!

Who gets into the college he wants to get into because he has good grades and consistent behavior? The person with the good brain or the person with the bad brain? The person with the good brain!

Who gets the job, has more money, has more lasting meaningful relationships, and lives longer because of the better decisions that are made throughout their lives? The person with the good brain or the person with the bad brain? The person with the good brain!

This exercise sort of takes the wind out of the sails of the teenager who wants to justify brain-hurtful behavior.

Here is a list of strategies and behaviors that decelerate brain aging. If you want to live a long time with your brain intact, engage in these behaviors:

- Make good decisions. Consistent, thoughtful, conscientious behavior is the number-one predictor of longevity!
- Surround yourself with a positive, healthy support system made up of friends and family.
- Protect your brain from injuries.
- Keep your surroundings free of toxins.
  - o Limit alcohol consumption to no more than four glasses a week.
  - o Protect yourself from excessive free radical formation.
- Seek healthy DNA repair mechanisms and ways to grow telomere length. Fish oil, taking a multiple vitamin, and drinking green tea have been associated with longer telomere length.
- Maintain physical health.
  - o Strive for low levels of inflammation.
  - o Avoid gum disease and intestinal problems.
  - o Maintain healthy levels of thyroid, testosterone, and other essential hormones.
  - o Maintain healthy levels of nutrients, such as vitamin D and omega-3s.
  - o Get physical exercise, and include endurance and strength training.
  - o Get quality sleep, seven to eight hours each night.
  - o Maintain a healthy weight and expend calories wisely.

- Focus on great nutrition that serves your brain and body.
  - o Eat high-quality calories and not too many of them.
  - o Drink plenty of water and avoid liquid calories.
  - o Eat high-quality lean protein.
  - o Eat "smart" (low-glycemic, high-fiber) carbohydrates.
  - o Limit fat consumption to healthy fats, especially those containing omega-3s.
  - o Eat natural foods of many different colors to boost antioxidants.
  - o Cook with brain healthy herbs and spices.
  - o Avoid sugar.
- Engage in physical exercise, building endurance and strength.
- Continue lifelong learning.
- Practice effective stress management practices, such as deep breathing and meditation.
- Promote good mental health and avoid severe anxiety or depression.
  - o Maintain a healthy level of anxiety to keep your behavior on track.
  - o Maintain an optimistic mood.
  - o Promote positive messages around aging.
- Pursue brain rehabilitation strategies when needed, such as neurofeedback and hyperbaric oxygen therapy.
- Get proper nutritional supplementation.
  - o Take a multivitamin.
  - o Take an omega-3 supplement.
  - o Take a vitamin D supplement.
  - o Consider taking individualized supplements to fit your brain type.
- Protect and repair your DNA—controlling oxidation and inflammation and avoiding toxins are all part of protecting your DNA, plus nutrients like green tea, omega 3s, multiple vitamins,

and superfoods such as grains like chia and quinoa; seaweeds and algae; and many spices may help.

- Develop meaning and purpose in your life.
- Know the state of the health of your brain.

6. **How do you know unless you look? Brain SPECT imaging is an essential tool at the Amen Clinics to help us understand the current state of our patients' brain function and to help us target the most appropriate interventions.**

From the time I ordered my first brain SPECT scan in 1991, I knew it was a powerful tool, because when people saw the scans it changed their behavior.

A sign of intelligent life is changing your behavior after receiving new information.

The SPECT scans are extraordinarily useful in giving us more information to help our patients who suffer with dementia, memory problems, brain injuries, depression, obsessiveness, substance abuse, attention deficit hyperactivity disorder (ADHD), anger issues, and more. When used in combination with detailed clinical histories, the scans help us target treatment to our patients' own specific brain patterns. Additionally, the scans help us use and develop more natural treatments to optimize the brain, in part because some of the typical pharmaceutical interventions appeared toxic on scans. One of the most important uses of the scans is that they help patients develop their own intimate relationship with their brains and be more willing to take better care of them.

Not all of my colleagues have embraced the clinical use of brain SPECT imaging in clinical practice, but more and more are doing so every day. Routinely, we get inquiries from physicians around the world who want to incorporate brain SPECT and these concepts into their practices. Without brain imaging tools to guide us, physicians wander around in the dark wondering what to really do to help their patients and themselves live long and stay young. In chapter 10 there will be more information on how brain SPECT imaging can be useful to you, even if you never get a scan.

7. **You are not stuck with the brain you have. You can make it better, even if you have been bad to it. You can change your brain and change your life.** This is one of the most exciting breakthroughs in medicine. This hopeful message has driven our work since 1991. At the Amen Clinics we have done thousands of before-and-after scans, like with Carlos, and have demonstrated that with targeted, "brain smart" interventions *you can boost the actual physical functioning of your brain and decrease its functional age, even if you have previously been bad to your brain!*

Intuitively, most people know that bad behaviors make us age more quickly. We see it in the skin of smokers or in the gaunt appearance of a methamphetamine addict or in the decreased cognitive functioning of people suffering from alcoholism. Unfortunately, in my experience, most people do not have a clue about how their physical health affects their cognitive and mental health.

For example, I did a fascinating project with a group of businesswomen. When your brain works right, your business works right! One of our CEOs, Tina, was struggling with depression, obesity, and uncontrolled diabetes. Her brain SPECT scan looked awful.

Tina told me that she had been diagnosed with diabetes several years earlier but had not found the time to really get healthy. She thought she would eventually get around to it. I thought to myself, "This is really nuts." (My prefrontal cortex, or PFC, in the front third of my brain, prevented me from saying it out loud.) Tina, like most people, had no idea that diabetes causes brain damage. It damages blood vessels, including those to the brain, and doubles the risk for Alzheimer's disease. Obesity, all by itself, is also a risk for thirty medical illnesses, including Alzheimer's disease; and, as we will see, it damages the brain. Depression can be caused by uncontrolled diabetes and obesity and is, by itself, another independent risk factor for Alzheimer's disease.

I looked at Tina and said, "Nothing in your life will work right, especially not your business, if your brain isn't right. Your physical

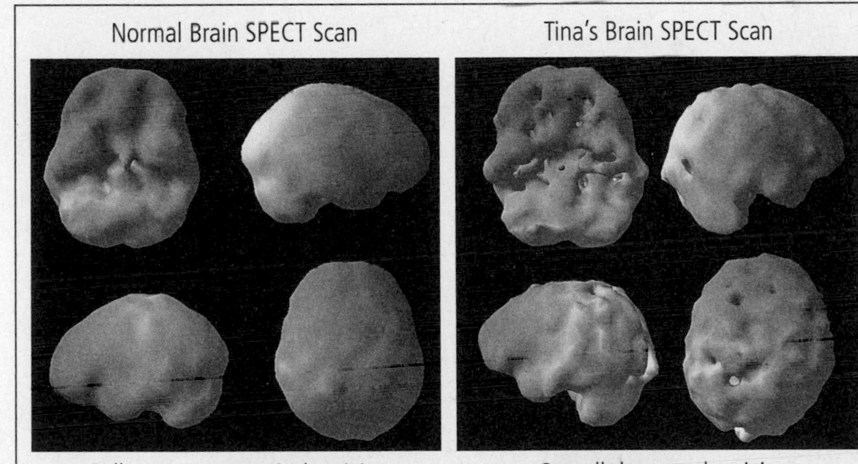

Full, even, symmetrical activity

Overall decreased activity

Tina's Follow-up SPECT Scan

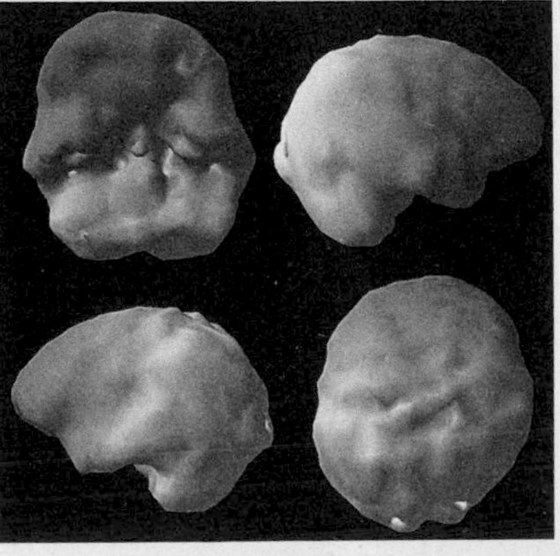

Overall increased activity

health problems are a 'brain emergency.' It is critical to get your weight and diabetes under control, and when you do this, your mood will be better as well."

Since we started working together, Tina has lost 40 pounds, her diabetes is under control, her mood is better, and she looks and feels much younger. Moreover, her business has dramatically improved because she has better focus, energy, and judgment.

NONE OF THIS IS HARD OR ABOUT DEPRIVATION! IT IS ABOUT SMART CHOICES AND GOOD DECISIONS

You will find that none of the strategies in this book are hard to do or require deprivation. The program is about making smart choices and good decisions. Mind-set here is critical. If you think something is being taken from you, you will resist. If you think you are being given the gift of extended, prosperous life for your mind and body, it will be easy to stay on this program. In fact, with the proper attitude, you will start to vigilantly protect your health from all those people in your life who are actively trying to steal it.

At this stage of my life, I am only interested in habits and food that serve me rather than those that enslave and steal from me. For example, I used to love Rocky Road ice cream until I learned that not only is it filled with sugar and calories, which promote obesity, inflammation, and erratic brain cell firing, it also has a type of fat, palmitic acid, that fools the brain into thinking it has not eaten anything at all. No wonder that halfway through the first bowl of ice cream, I was thinking about the second bowl. This does not happen with nonfat yogurt and blueberries, which I also love.

Success on this program will require you to use the PFC to plan and think ahead. Strengthening this part of the brain will dramatically improve your focus, forethought, judgment, and impulse control. Throughout the book I will give you hundreds of ways to enhance your

PFC and improve every decision you make in your life to keep you on track toward your goals. Success will also require a powerful emotional or limbic brain to keep you motivated to move in the direction of health.

WHY YOU WANT TO COMPLETELY EMBRACE THIS PROGRAM *NOW*

By adopting the brain healthy strategies detailed in this book, you can outsmart your genes, put the brakes on aging, and even reverse the aging process so that you look and feel younger in a very short period of time. Researchers indicate that within just three months of adding a new healthy habit, you can start seeing a measurable difference in your life expectancy. By following the plan outlined in this book, your brain can start looking younger in a matter of just eight weeks.

In my lectures I often ask the audience, "How many of you want to live until eighty-five or beyond?" Most of the audience raises their hands. "Did you know," I continue, "that 50 percent of people eighty-five or older will be diagnosed or have significant symptoms of Alzheimer's disease or other forms of dementia?" That one statistic gets their attention and it should get yours as well.

How is Alzheimer's disease different from dementia? Dementia is the big umbrella category. Alzheimer's is one of the common types of dementia. Other types of dementia include alcoholic dementia, brain trauma dementia, vascular dementia (often associated with small or large strokes), pseudodementia (depression that mimics dementia), and frontal temporal lobe dementia, to name a few.

With the aging population, Alzheimer's disease is expected to triple from five million to fifteen million Americans in the coming decades, and there is no cure for it on the horizon. If that is not enough motivation for you to get healthy, then you might want to volunteer at an elderly care center for two weeks and meet a few people with Alzheimer's disease and other forms of dementia. It is a frightening illness that robs you of your ability to form new memories. Later in the illness, you lose old memories too. And it puts a tremendous burden on families.

One of the reasons there is not likely to be a cure for Alzheimer's disease and other forms of dementia is that they start thirty years or more before people have any symptoms. According to one study at UCLA, 95 percent of people with Alzheimer's disease were diagnosed when they were in the moderate to severe stages of the illness, when not much can be done. Early screening and intervention is absolutely essential.

The National Institute of Aging recently revised its staging guidelines for Alzheimer's disease. The old guidelines had three stages:

1. Normal, in which people had no symptoms
2. Mild cognitive impairment, in which people or relatives started to notice a problem
3. Alzheimer's disease, in which a significant problem was present

Based on new brain imaging data, the National Institutes of Aging added a new stage:

1. Normal
2. Preclinical stage, in which there were no obvious outward symptoms but where negative changes were already brewing in the brain
3. Mild cognitive impairment
4. Alzheimer's disease

Can you see the problem here? You have no symptoms at all, but your brain is already starting to dramatically deteriorate, up to thirty to fifty years before you have symptoms! The time to start preventing Alzheimer's and other diseases of aging is now, not tomorrow, no matter what your age. The person who is diagnosed with Alzheimer's disease at age fifty-nine likely started to show noxious brain changes by age thirty. The person who is diagnosed with Alzheimer's disease in his early seventies already had evidence of brain deterioration in his forties.

Losing your memory or developing brain fog in your forties, fifties, sixties, seventies, or even eighties is not normal. It is a sign of trouble. Be smart and stop waiting for a problem to hit you in the head before you decide you have to get healthy.

Marianne was fifty-nine years old when she almost quit her job. She worked as a high-level business executive but felt as though her mind was beginning to deteriorate. Physically, her whole body hurt, and her head felt foggy all day long. At first she thought that she was "just" getting older, that it happened to everyone. But as she got worse, she thought it was unfair to her co-workers that she was not on top of her game, and she considered leaving a job she loved. She thought her best days were behind her. Her daughter gave her a copy of one of my books and she immediately started the program. To her amazement, within two months she started to feel much better, her pain was gone, and the brain fog had lifted. Within a year, she had lost thirty pounds and now her brain feels younger, sharper, and more energized than it has in decades. "I have a fast-acting brain with the wisdom of experience," she told me. "I feel like I am at the peak of my life and my best is no longer behind me."

Are the treatments we have now for Alzheimer's disease more likely to make an impact early or late in the illness? Early. The more brain tissue there is available to save, the better. The best tactic to decrease your risk for Alzheimer's disease or to even prevent it is to decrease the illnesses and problems that are associated with it, such as brain injuries, drug or alcohol abuse, heart disease, vascular diseases, strokes, cancer, obesity, sleep apnea, diabetes, hypertension, depression, toxic exposure, low testosterone levels, and low thyroid levels.

This book will be your road map for decreasing your risk for Alzheimer's disease and other forms of dementia, and in the process you will look and feel better, have a sharper memory and improved decision-making skills. Since the problems of aging, including dementia, start much earlier than their symptoms manifest, *now* is the time to take your brain health seriously, no matter what your age. Many of the diseases of aging actually start in childhood or adolescence and include obesity, brain injuries, depression, and an unhealthy support system. Keeping your children and grandchildren healthy is one of the best gifts you can give them. Knowing what we know now, if you are the grandparent who gleefully doles out the candy and ice cream whenever the grandchildren come over, you are clearly not doing them any favors.

This book is organized around ten stories that highlight the major concepts of using your brain to change your age.

1. Nana, Lisa, and Ruth's story will show you the importance of knowing your important health numbers. You cannot change what you do not measure. Allowing these numbers to get out of control may very well lead to your early death. *It will also give you a plan to decrease your risk for Alzheimer's disease and age-related memory problems.*

2. Tamara's story will show you how the food you eat literally poisons your body, drains your brain, and causes you to want to check out of life early, or how it can be your best medicine. If you are smart, you will only want to eat food that serves you.

3. Andy's story will show you the importance of getting strong to live long. His before-and-after brain scans will show you how you can look and feel dramatically younger by following the steps of the program, especially adding consistent physical exercise.

4. Jose's story directly demonstrates how the physical health of your brain enhances good decision-making skills to help you live longer.

5. Jim's story highlights the importance of lifelong learning to keep the brain young.

6. Joni's story illustrates the connection between brain health, beautiful skin, and a healthy sex life. The health of your skin is an outside reflection of the health of your brain. Since the brain is 50 percent visual, healthy skin attracts others to you.

7. Chris and Sammie's heartbreaking story will demonstrate the need to treat grief, depression, anxiety disorders, and other emotional challenges in order to want to live long and feel younger.

8. Anthony, Patrick, Nancy, and others will demonstrate that brain damage is often reversible with an intense, smart, focused program.

9. "The Tale of Two Ricks" will show the importance of your "people" connections to get and stay healthy.

10. Daniel—that's me—and Brain SPECT Imaging will tell how the brain images in this book have changed everything in my life and how they can change yours as well, even if you never get a scan.

    In addition, there will be detailed information on ways to actively protect the brain from aging, as well as on the smart use of natural supplements. I am excited to be your guide on this journey. Together we can make a difference in your life and in the lives of those you love, even for three or four generations.

STEVE'S AMAZING TRANSFORMATION

Eighteen months before I wrote this book Steve was forty years old and weighed 630 pounds. He was depressed and struggling with a twenty-five-year addiction to alcohol and nicotine. He suffered with sleep apnea, hypertension, diabetes, and severe chronic pain in his feet,

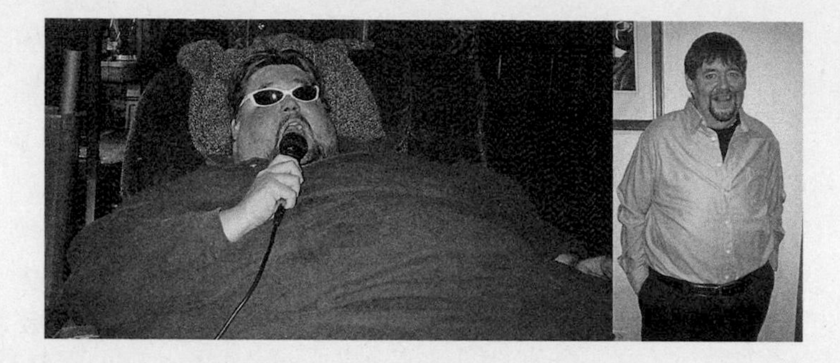

which tortured him day and night. He was so big that whenever he fell he would have to call 911 for a team of people to help him. At the time, he was contemplating suicide and finally decided that he had only two choices: either to live or to die. He chose life.

His sister bought him a copy of my book *Change Your Brain, Change Your Body,* which he followed religiously, and together with the help of many supportive people, Steve lost 156 pounds over the next four months and has now lost over 380 pounds. In addition, Steve has gone from ten medications to two, and he has lost his pain, diabetes, cigarettes, and depression—all without any surgery. Not only does Steve look and feel dramatically younger, but his brain is younger as well. He has better focus, better energy, and a better memory. Ultimately, Steve used his brain to change his age, and in the process he saved his life. If Steve can use these principles to get healthy, then I know you can too.

1

NANA, LISA, AND RUTH

KNOW YOUR NUMBERS TO KEEP YOUR MIND HEALTHY AND PREVENT ALZHEIMER'S AND OTHER DISEASES OF AGING

I can never find my keys. Sometimes they show up by the eggs in the refrigerator.
I am fifty-two. Isn't that normal?
Think again!

When Lisa was a young girl she adored her nana, her mother's mother. Nana and Lisa baked cookies together, played cards for hours, told silly jokes, and picked plums in Nana's backyard. Nana taught Lisa how to can the fruit for plum jam, which they loved to share. Nana was very overweight, so she would hold the ladder while her granddaughter climbed the ladder for the plums. On nights Lisa slept over, Nana always read to her. Lisa remembers laughing so hard that she would sometimes snort at the silly voices Nana used when she read the stories. At night in the dark they promised each other to always be best friends. Lisa loved snuggling into Nana's body, which was ever so soft. She felt unconditional love in Nana's presence, which was one of the best feelings she remembered from her childhood.

Then, when Lisa was about twelve years old, something started to change. At first, it was barely noticeable. Nana seemed less interested in their time together. There were no more jokes, fewer stories, and Nana said she was too tired to play games or pick plums. Nana was also more irritable with Lisa, even sometimes yelling at her for what seemed like no reason at all. Lisa was devastated, but Nana did not pick up on the social cues that should have told her that her granddaughter needed soothing. Lisa remembers this as one of the saddest, most confusing times in her life. She wondered if she had done something to make Nana mad. "What's wrong with Nana?" Lisa would ask her mother, but time and again her mother would say, "Don't worry. Nana is fine." This only deepened Lisa's pain and confusion. Maybe she really was the problem and Nana had just stopped loving her.

Her grandmother was sixty-five years old when Lisa noticed the changes. Around this time, Nana had been diagnosed with diabetes and high blood pressure. Lisa remembered watching Nana take her pills and her shots to feel better, but no one seemed overly concerned about her health.

When Lisa was fourteen, Nana took a dramatic turn for the worse. With Lisa in the car, Nana got lost on the way home from the store. Nana panicked and stopped a man who was walking across the street to ask for help, but she could not tell him where she lived. She appeared frightened and confused, like a child. Lisa asked the man to call her grandfather, who came to pick them up.

Once they got home Lisa cornered her mother. "Look, Mom, I know something is really wrong with Nana. Her brain isn't working right. She needs help." Still, the family continued making excuses, normalizing what was obviously not normal behavior. Looking back on this time, as an adult, Lisa remembers being furious, feeling she was, even as a young teen, the lone voice of reason shouting into a bitter wind. After Nana got lost several more times, the family finally was concerned enough to take her to a doctor who diagnosed her with something called senile dementia. He recommended Nana live in a nursing home for people with memory problems.

Gone were the happy warm feelings she once enjoyed when she

visited her grandmother. The nursing home where she now lived smelled "medical" and felt cold, and Lisa felt odd and afraid in it. She never knew which Nana she'd find on these visits: Sometimes Nana smiled when she saw Lisa; sometimes she did not recognize her at all. Sometimes when Lisa read to Nana she seemed engaged and happy, other times her grandmother just wanted to be left alone. After a few years, Nana died in the nursing home. However, Lisa felt that Nana had really died years earlier when her personality slowly ebbed away. At Nana's funeral, all of their special times circled through Lisa's mind. She couldn't help wondering how a person could disappear while her body continued living on, and she couldn't help feeling how sad it all was. Lisa wondered if she or her mother would have the same problem as Nana. She prayed to God they would not.

Lisa's mother, Ruth, was also a lot of fun. They too had many special times, cooking, reading, and playing together. Like Nana, Ruth was a fabulous baker who also struggled with her weight, early onset diabetes, and hypertension. Lisa's mother was also a wonderful grandmother to Lisa's three daughters, which reminded Lisa of the closeness she'd shared with her own nana. In fact, her girls called her mother Nana as well. In the back of her mind she kept watch over her own mother's brain health. She didn't want her granddaughters to lose this vibrant and wonderful relationship they'd enjoyed with her mom, as she'd lost hers with Nana. It was this concern that prompted Lisa, now in her early forties, to pick up my book, *Change Your Brain, Change Your Life.*

When Ruth turned sixty-eight Lisa's worst fears started to actualize. At first, Ruth struggled with finding the right words. If she meant *dog,* she might unintentionally say *bark;* if she meant *milk,* she sometimes said *cow.* One time when she asked her granddaughter for a hug, she said, "Give Nana a slap."

Ruth's memory was also becoming a problem. Lisa watched her reach for the phone to dial her sister whom she'd just called five minutes earlier. Her sister said this sort of thing was happening more frequently.

Lisa's father mentioned that there were times when he found her mother standing and staring and not knowing why she was in a room. There were also two occasions on which Ruth got lost driving in a town where she had lived for thirty years, forcing her to call her husband for directions. Her father had installed a GPS system in Ruth's car to help her. (I sometimes wonder if having GPS systems actually delays the diagnosis of early Alzheimer's disease, as people do not have to rely as heavily on their own memories to get from point A to point B, so their deficits are not seen early by those who could encourage them to get help.)

Initially, Lisa's dad just laughed off her mother's struggles. He explained it away with, "She is just getting older. She's under a lot of stress." Or "You know your mom has never had a good memory or sense of direction. It will pass. Everything is all right."

Because early signs of dementia may alternate with periods of lucidity, families tend to deny what's happening. This is tragic, because the earlier someone seeks help, the better the prognosis. Remembering her Nana, Lisa wasn't about to ignore her concerns or let others downplay them. She anxiously and emphatically told her father, "Mom needs help and she needs it now." Together they approached her mom with their observations and concerns, urging her to go to the Amen Clinics. At first Ruth was resistant. "I'll be okay," she said, which frightened Lisa even more. Then Lisa reminded her of Nana and told her that early intervention might help her avoid Nana's fate. At this, Ruth agreed to come to the Amen Clinics for an evaluation and brain SPECT imaging.

I greeted Lisa and Ruth when they came into my office, and listened to their story. From these descriptions alone, I suspected Ruth had early Alzheimer's disease. However, after being a neuropsychiatrist for thirty years, I knew I couldn't proffer a diagnosis based on my suspicions alone. I had to look, test, probe, and get as much information as possible.

Ruth's brain SPECT scan showed three findings consistent with Alzheimer's disease:

1. Decreased activity in her parietal lobes, at the back, top part of her brain. The parietal lobes help with direction sense.

2. Decreased activity in her temporal lobes, which help get memories into long-term storage. The temporal lobes are also involved in word finding.

3. Decreased activity in an area called the posterior cingulate gyrus, deep in the back part of the brain. The posterior cingulate gyrus is one of the first areas in the brain that dies in Alzheimer's disease and is involved with visual memory.

   The structured memory testing that we do at the Amen Clinics also showed significant problems in both immediate and delayed recall.

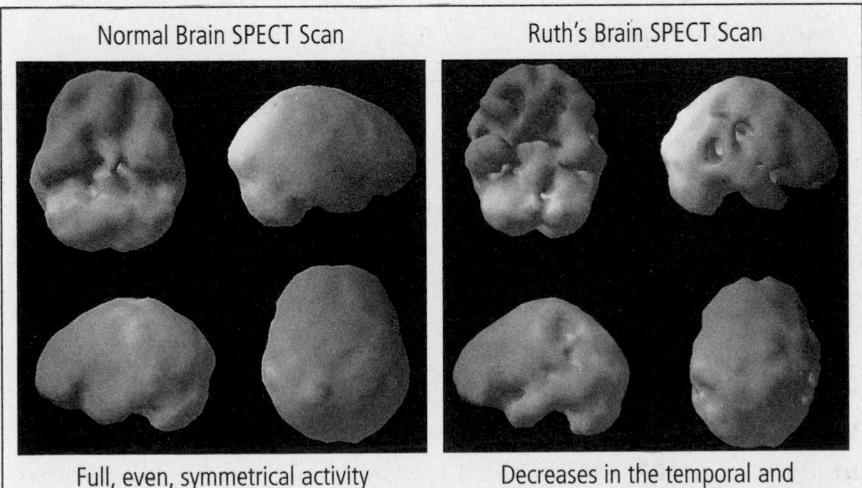

Full, even, symmetrical activity

Decreases in the temporal and parietal lobe consistent with early Alzheimer's disease

KNOW YOUR NUMBERS: YOU CANNOT CHANGE WHAT YOU DO NOT MEASURE

The next step in the process was to look at her important numbers. These are also vital numbers for you to know and optimize; they're a critical step in helping you live longer and look younger.

Here is a list we use at the Amen Clinics.

1. BMI
2. Waist-to-height ratio
3. Calories needed / calories spent
4. Number of fruits and vegetables eaten a day
5. Average of hours slept each night (with sleep apnea ruled out)
6. Blood pressure
7. Complete blood count
8. General metabolic panel with fasting blood sugar
9. HgA1C test for diabetes
10. Vitamin D level
11. Thyroid levels
12. C-reactive protein
13. Homocysteine
14. Ferritin
15. Testosterone
16. Lipid panel
17. Folic acid and B_{12} levels
18. Syphilis and HIV screenings
19. Apolipoprotein E genotype test
20. Twelve modifiable health risk factors

1. **Know Your BMI** Ruth's BMI, or body mass index, was 32, which was too high. A normal BMI is between 18.5 and 24.9, overweight

is between 25 and 29.9, and obese is greater than 30. You can find a simple BMI calculator on our website at www.amenclinics.com. Ruth was obese, which was not helping her brain remain healthy. As already mentioned, being obese has been associated with less brain tissue and lower brain activity. Obesity doubles the risk for Alzheimer's disease. There are probably several mechanisms that underlie this finding, including the fact that fat cells produce inflammatory chemicals and store toxic materials in the body.

One of the reasons I want my patients to know their BMI is that it stops them from lying to themselves about their weight. I was sitting at dinner recently with a friend who seemed totally indifferent about his weight, even though he was injecting himself with insulin for his diabetes at the table. As we were talking, I calculated his BMI for him. Trust me, I can be a very irritating friend if I think you are not taking care of yourself. His BMI was just over 30, in the obese range. That really got his attention. Since then he has lost 20 pounds and is more committed to getting healthy. The truth will set you free. Know your BMI.

I put Ruth on a structured weight-loss program.

2. **Know Your Waist-to-Height Ratio (WHtR)** Another way to measure the health of your weight is with your waist-to-height ratio. Some researchers believe this number is even more accurate than your BMI. BMI does not take into account an individual's frame, gender, or the amount of muscle mass versus fat mass. For example, two people can have the same BMI, even if one is much more muscular and carrying far less abdominal fat than the other; this is because BMI does not account for differences in fat distribution. The WHtR is calculated by dividing waist size by height. As an example, a male with a 32-inch waist who is 5'10" (70 inches) would divide 32 by 70 to get a WHtR of 45.7 percent. The WHtR is thought to give a more accurate assessment of health, since the

most dangerous place to carry weight is in the abdomen. Fat in the abdomen, which is associated with a larger waist, is metabolically active and produces various hormones that can cause harmful effects, such as diabetes, elevated blood pressure, and altered lipid (blood fat) levels. Many athletes, both male and female, who often have a higher percentage of muscle and a lower percentage of body fat, have relatively high BMIs, but their WHtRs are within a healthy range. This also holds true for women who have a "pear" rather than an "apple" shape.

You want your waist size in inches to be less than half your height. So if you are 66 inches tall, your waist should not be more than 33 inches. If you are 72 inches tall, your waist should not be more than 36 inches.

Note: You have to actually measure your waist size with a tape measure! Going by your pants size does not count, as many clothing manufacturers actually make their sizes larger than they state on the label so as not to offend their customers. I remember preferring to purchase pants or shorts that were labeled "relaxed fit" because I could still fit into a size 34-inch waist. There was no way I could get into a slim-fit 34 inches, which, looking back, actually was 34 inches. Since doing this work, I have seen that most people do not know their waist size and are in total denial. Most of our NFL players and patients significantly underestimate their waist size. One of the pastors we work with said his waist size was 42 inches, but when we measured (at the belly button) it was really 48 inches. Ruth was 5'4" tall. She told me her waist was 33 inches. It measured at 37 inches.

This was further confirmation that Ruth needed a structured weight-loss program.

3. **Know the Number of Calories You Need and Spend in a Day** I think of calories like money: If I eat more than I need, my body

will become bankrupt. Wise caloric spending is a critical component to getting healthy. Don't let anyone tell you that calories don't count. They absolutely do. The people who say calories don't matter are just fooling themselves. You need to know how many calories you need to eat a day to maintain your current weight. The average fifty-year-old woman needs about 1,800 calories, and the average fifty-year-old man needs about 2,200 calories a day. This number can go up or down based on exercise level and height. You can find a free personalized "caloric need" calculator at www.amenclinics.com.

Set a realistic goal for your desired weight and match your behavior to reach it. If you wish to lose a pound a week, you typically need to eat 500 calories a day *fewer* than you burn. I am not a fan of rapid weight loss. It does not teach you how to live for the long term. One of my patients went on the hCG diet and lost 40 pounds in three twenty-six-day cycles, but it was at a pretty high cost. Within the next six months, she put all the weight back plus another 10 pounds. Slow and steady teaches you new habits. I like it for people to lose a pound a week, which teaches them a new way of living for the long term.

Know the Daily Calories You Consume Stomp out calorie amnesia! For anyone who has problem with their weight, this is a very effective strategy to get back on track. Stop lying to yourself about what you are actually putting into your body. As we will discuss in the next chapter, think CROND: calorie restricted, optimally nutritious, and delicious. Besides becoming familiar with calorie counts, get a small notebook that you can carry with you everywhere. It will be your new best friend. Jot down the calories you consume as you eat during the day. If you keep this log, together with the other parts of the program, it will be a major step forward in getting control of your brain and body for the rest of your life. If you don't know the calories of something, don't eat it. Why are you going to let someone else sabotage your health? *Ignorance is not bliss. It increases your chances for an early death.*

Until you really understand calories, you need to learn to weigh and measure food and look at the food labels for portion size. What the cereal companies think is a portion size may not be anywhere near what your eyes think. When you actually do this, I can promise it will be a rude awakening. I know it was for me. Upon keeping track of his calories, one of our NFL players wrote, "I had no idea of the self-abuse I was doing to my body!"

Ruth had no idea of how many calories she needed a day or how many she actually ate. This had to be part of her brain rehabilitation program.

4. **Know the Number of Fruits and Vegetables a Day You Eat** Count them! Eat more vegetables than fruits and try to get that number to between five and ten servings to enhance your brain and lower your risk for cancer. Ruth said she was erratic in eating vegetables and had no idea of the actual number of servings she was eating each day. Another benefit of eating five to ten servings a day of fruit and vegetables is that they are so naturally filling, making it much easier to keep within your calorie limits. Ruth's diet needed a makeover.

5. **Know How Many Hours of Sleep You Get Each Night** Ruth typically got five hours of sleep at night. Her husband said she did not snore or stop breathing at night. Sleep assessment with memory problems and aging is critical. One of the fastest ways to age is by getting less than seven or eight hours of sleep at night. People who typically get six hours of sleep or less have lower overall blood flow to the brain, which hurts its function. Researchers from the Walter

Reed Army Institute of Research and the University of Pennsylvania found that chronically getting less than eight hours of sleep was associated with cognitive decline.

Chronic insomnia triples your risk of death from all causes.

This was clearly an area in which Ruth needed help. I recommended sleep strategies, including a warm bath before bed, no television an hour before bed, a sleep-inducing hypnosis CD, and a melatonin-based sleep supplement.

While we are discussing sleep, it is important to know that sleep apnea doubles a person's risk for Alzheimer's disease. On our brain SPECT studies sleep apnea often looks like early Alzheimer's disease with low activity in the parietal and temporal lobes. Sleep apnea is characterized by snoring, periods of apnea (temporary cessation of breathing), and chronic tiredness during the day. The chronic lack of oxygen from the apnea periods is associated with brain damage and early aging. Sleep apnea has also been associated with obesity, hypertension, strokes, and heart disease. If there is any chance you may have sleep apnea, go to your health care professional who can refer you to a sleep lab.

6. **Know Your Blood Pressure** Ruth's blood pressure was 145/92 mm/Hg (millimeters of mercury) on her blood pressure medication. This is way too high. High blood pressure is associated with lower overall brain function, which means more bad decisions. Check your blood pressure or have your doctor check it on a regular basis. If your blood pressure is high make sure to take it seriously. Some behaviors that can help lower your blood pressure include losing weight, daily exercise, fish oil and, if needed, medication.

Optimal: Below 120/80 mm/Hg

Prehypertension: 120/80 to 130/80–130/89 mm/Hg

Hypertension: 140/90 mm/Hg or above

I added exercise and high-dose fish oil to Ruth's regimen.

LABORATORY TESTS

The next set of important numbers comes from laboratory tests that are usually ordered by a health care professional. It is essential that you know these important numbers about yourself.

7. **Know Your Complete Blood Count (CBC)** You need to check the health of your blood, including red and white blood cells. People with low blood count can feel anxious, tired, and have significant memory problems. In one of our patients who was screened as part of a regular physical, we picked up that he had leukemia, even though he had no physical symptoms. Early treatment for most medical conditions, including leukemia or Alzheimer's disease, is best. Ruth's CBC was normal.

8. **Know Your General Metabolic Panel** This is to check the health of your liver, kidneys, fasting blood sugar, and cholesterol. Ruth's fasting blood sugar was high at 135. Normal is 70–99 mg/dL (milligrams per deciliter), prediabetes is 100–125 mg/dL, and diabetes is 126 mg/dL or higher. Even though Ruth was being treated, her blood sugar was too high.

   Why is high fasting blood sugar a problem? High blood sugar causes vascular problems throughout your whole body. It causes blood vessels over time to become brittle and vulnerable to breakage. It leads not only to diabetes but also heart disease, strokes, visual impairment, impaired wound healing, wrinkled skin, and cognitive problems.

If we were going to reverse Ruth's cognitive deterioration, it was critical to get her blood sugar under better control with a healthy diet and some simple supplements, such as alpha-lipoic acid.

9. **Know Your HgA1C Level** This test shows your average blood sugar levels over the past two to three months and is used to diagnose diabetes and prediabetes. Normal results for a nondiabetic person are in the 4–6 percent range. Prediabetes is indicated by levels in the 5.7–6.4 percent range. Numbers higher than that may indicate diabetes.

   Ruth's HgA1C was high at 7.4 percent. To optimize it, I recommended that she lose weight, eliminate all sugar and refined carbohydrates, eat several small meals a day with some protein at each meal, exercise, and begin taking fish oil and the supplement alpha-lipoic acid.

10. **Know Your Vitamin D Level** Low levels of vitamin D have been associated with obesity, depression, cognitive impairment, heart disease, reduced immunity, cancer, psychosis, and all causes of mortality. Have your physician check your 25-hydroxy vitamin D level, and if it is low get more sunshine and/or take a vitamin D_3 supplement. A healthy vitamin D level is 30–100 ng/dL (nanograms per deciliter). Optimal is 50–100 ng/dL. Two-thirds of the U.S. population is low in vitamin D, the same number of Americans who are overweight or obese. One of the reasons for the dramatic rise in vitamin D deficiency is people wearing more sunscreen and spending more time inside working or in front of the television or computer.

    Ruth's vitamin D level was 8 ng/dL, which was very low. Optimizing her vitamin D level was another critical component to optimizing her cognitive health.

11. **Know Your Thyroid Levels** Abnormal thyroid hormone levels are a common cause of forgetfulness, confusion, lethargy, and other symptoms of dementia in both women and men. Having low thyroid levels decreases overall brain activity, which can impair your thinking, judgment, and self-control, and make it very hard for you to feel good. Low thyroid functioning can make it nearly impossible to manage weight effectively. Know your:

    - TSH (thyroid-stimulating hormone)—normal is between 0.350 and 3.0 μIU/mL
    - Free T3 (300–400 pg/dL [picogram per deciliter])
    - Free T4 (1.0–1.80 ng/dL)
    - Thyroid peroxidase (TPO) antibodies (0–34 IU/mL)

There is no one perfect way, no one symptom or test result, that will properly diagnose low thyroid function or hypothyroidism. The key is to look at your symptoms and your blood tests, and then decide. Symptoms of low thyroid include fatigue, depression, mental fog, dry skin, hair loss, especially outer third of eyebrows, feeling cold when others feel normal, constipation, hoarse voice, and weight gain.

Most doctors do not check TPO antibodies unless the TSH is high. This is a big mistake. Many people have autoimmunity against their thyroid, which makes it function poorly but still register a "normal" TSH. That's why I think this should be part of routine screening.

Medication can easily improve symptoms if a thyroid problem is present. Have your doctor check your thyroid hormones for hypothyroidism or hyperthyroidism and treat as necessary to normalize.

Ruth's thyroid tests were normal.

12. **Know Your C-reactive Protein Range** This is a measure of inflammation. Elevated inflammation is associated with a number of diseases and conditions that are associated with aging and cognitive impairment. Fat cells produce chemicals that increase inflammation. A healthy range is between 0.0 and 1.0 mg/dL. This is a very good test for inflammation. It measures the general level of inflammation but does not tell you where it is from. The most common reason for an elevated C-reactive protein is metabolic syndrome or insulin resistance. The second most common is some sort of reaction to food—either a sensitivity, a true allergy, or an autoimmune reaction as occurs with gluten. It can also indicate hidden infections.

    Ruth's C-reactive protein test was 7.3 mg/dL, which was way too high and needed to be addressed immediately with high-dose fish oil (6 grams a day) and the same healthy anti-inflammatory diet recommended in this book.

13. **Know Your Homocysteine Level** Elevated levels (>10 μmol/L [micromoles/liter]) in the blood have been associated with damage to the lining of arteries and atherosclerosis (hardening and narrowing of the arteries) as well as an increased risk of heart attacks, strokes, blood clot formation, and possibly Alzheimer's disease. This is a sensitive marker for B vitamin deficiency, including folic acid deficiency. Replacing these vitamins often helps return the homocysteine level to normal. Other possible causes of a high homocysteine level include low levels of thyroid hormone, kidney disease, psoriasis, some medicines, or when the condition runs in your family. The ideal level is 6–10 μmol/L (micromoles/liter). Eating more fruits and vegetables (especially leafy green vegetables) can help lower your homocysteine level by increasing how much folate you get in your diet. Good sources of folate include lentils, asparagus, spinach, and most beans. If adjusting your diet

is not enough to lower your homocysteine, take folic acid (1 mg), vitamin B_6 (10 mg), and B_{12} (500 μg [micrograms]).

Ruth's homocysteine level was high at 16 μmol/l. I recommended a comprehensive multivitamin with higher levels of B vitamins and a healthy diet.

14. **Know Your Ferritin Level** This is a measure of iron stores that increases with inflammation and insulin resistance. Less than 200 ng/mL is ideal. Women tend to have lower iron stores than men, due to blood loss (blood cells contain iron) from years of menstruation. Low ferritin levels are associated with anemia, restless legs, and ADD. Higher iron stores have been associated with stiffer blood vessels and vascular disease. Some research suggests that donating blood to lower high ferritin levels may enhance blood vessel flexibility and help decrease the risk of heart disease. Moreover, whenever you give blood you are being altruistic, which will also help you live longer.

    Ruth's ferritin level was normal.

15. **Know Your Free and Total Serum Testosterone Levels** Low levels of the hormone testosterone, for men or women, have been associated with low energy, cardiovascular disease, obesity, low libido, depression, and Alzheimer's disease.

    Normal levels for adult males are:

    - Testosterone Total Male (280–800 ng/dL)
    - Testosterone Free Male (7.2–24 pg/mL)

    Normal levels for adult females are:

    - Testosterone Total Female (6–82 ng/dL)
    - Testosterone Free Female (0.0–2.2 pg/mL)

Ruth's free and total testosterone levels were very low. Sometimes hormone replacement is necessary, but my first intervention is a healthy diet, which eliminates sugar. Getting a sugar burst has been associated with lower testosterone levels.

16. **Know Your Lipid Panel** Sixty percent of the solid weight of the brain is fat. High cholesterol is obviously bad for the brain, but having too little is also bad, as some cholesterol is essential for making sex hormones and helping the brain function properly. Getting your lipid panel checked regularly is important. This test includes HDL (high-density lipoprotein, or "good" cholesterol), LDL (low-density lipoprotein, or "bad" cholesterol), and triglycerides (a form of fat). According to the American Heart Association, optimal levels are as follows:

- Total Cholesterol (<200 mg/dL)
- HDL (≥ 60 mg/dL)
- LDL (<100 mg/dL)
- Triglycerides (<100 mg/dL)

If your lipids are off make sure to get your diet under control as well as take fish oil and exercise. Of course you should see your physician. Also, knowing the particle size of LDL cholesterol is very important. Large particles are less toxic than smaller particle size.

Ruth's total cholesterol and LDL were high while her HDL was low.

17. **Know Your Folic Acid and B_{12} Levels** It is important in evaluating memory problems to rule out deficiencies of these nutrients. Ruth's levels were normal. I once had a patient who had a severe B_{12} deficiency, whose brain SPECT scan showed severe overall decreased blood flow.

18. **Know Your Syphilis and HIV Screening Results** Dementia can be associated with later-stage syphilis and HIV infections. If the person had syphilis or an HIV infection many years ago and was never properly treated, the illness may have progressed to the point of affecting behavior and intelligence. Even though it wasn't likely in Ruth's case, it is always important to check. Her tests were negative.

19. **Know Your Apolipoprotein E (APOE) Genotype** This test checks genetic risk. The presence of the APOE e4 gene significantly increases a person's risk for Alzheimer's disease and is associated with symptoms that appear five to ten years earlier than in the general population. Many children of an affected parent wish to know the parent's APOE genotype so they can determine their chance of inheriting a higher risk for AD, atherosclerosis, heart disease, and stroke.

    Everyone has two APOE genes, and if one of them—or worse, two of them—are APOE e4, that person's chances of having memory issues are higher. APOE genes alone are not dangerous; we need them to function. They help in the development, maturation, and repair of cell membranes of neurons, and they help regulate the amount of cholesterol and triglycerides in nerve cell membranes. There are three versions of the APOE gene: e2, e3, and e4, and it is the last one that is the culprit. As with all genes, we inherit one copy from each parent, and any one person could have the following combination: e2/e2, e2/e3, e2/e4, e3/e3, e3/e4, or e4/e4.

    If a person has two e4 genes, it means he received one from each parent. Because the APOE e4 gene is known to increase the beta-amyloid deposition and plaque formation that is found in people with Alzheimer's disease, it increases the chance of developing the most common form—late-onset Alzheimer's disease—by

2.5 (for one e4) or fivefold (for two e4s). The APOE e4 gene also causes symptoms to appear two to five years earlier than for those who don't have it but do have some other cause of Alzheimer's.

For about 15 percent of the general population, at least one of their two APOE genes is the e4 gene. People who have no APOE e4 gene at all have only a 5–10 percent chance of developing AD after age sixty-five, whereas people with one APOE e4 gene have about a 25 percent chance. That is quite a jump. But the good news that can be inferred here is that not everyone with the gene will develop Alzheimer's; in fact, 75 percent will not. One other thing to consider is that even if a person has one APOE e4 gene and he develops dementia, Alzheimer's disease might not be the source. There is a chance that the cause of the dementia could be something else. If the person has two APOE e4 genes, on the other hand, and he develops dementia, the odds are very good that it is from Alzheimer's disease. In fact, the odds are 99 percent.

Ruth had the e3/e4 gene.

20. Know How Many of the Twelve Most Important Modifiable Health Risk Factors You Have, Then Work to Decrease Them

Here is a list from researchers at the Harvard School of Public Health. Circle the ones that apply to you.

- Smoking
- *High blood pressure*
- *BMI indicating overweight or obese*
- *Physical inactivity*
- *High fasting blood glucose*
- *High LDL cholesterol*
- Alcohol abuse (accidents, injuries, violence, cirrhosis, liver disease, cancer, stroke, heart disease, hypertension)
- *Low omega-3 fatty acids*

- *High dietary saturated fat intake*
- *Low polyunsaturated fat intake*
- *High dietary salt*
- *Low intake of fruits and vegetables*

Ruth had ten of the twelve preventable risk factors (those in italics) for early death. Working to address these issues was critical to have any hope of reversing the negative trend.

In summary, Ruth's evaluation demonstrated a clinical picture, scan findings, and memory testing consistent with "early" Alzheimer's disease. She had one of the APOE e4 genes that put her at risk, and she was practicing only two of the twelve preventable risk factors for early death. The good news in a case such as Ruth's was that there were so many different important numbers that could still be changed or optimized that it might make a significant difference in her mental state.

Thankfully, Ruth was still cognitively aware enough and still had ample prefrontal cortex function to understand the severity of her problems were likely to get worse without a serious attempt at getting well. The plan for Ruth included these elements:

- Immediately change to a CROND diet, eliminating sugar, simple carbohydrates, and artificial sweeteners and eating more vegetables (at least five servings) together with lean protein and healthy fat. She was to weigh herself every day, and have her husband record it for her.
- Begin taking supplements of multivitamins with extra B vitamins, alpha-lipoic acid for blood sugar regulation, vitamin D, and fish oil.
- Begin taking supplements to enhance memory, including vinpocetine and ginkgo to enhance blood flow, huperzine A and acetyl-L-carnitine to enhance the neurotransmitter

acetylcholine, and N-acetylcysteine (NAC), which is a super-antioxidant.

- Focus on getting seven to eight hours of sleep at night with better sleep habits, self-hypnosis, and melatonin as needed.
- Implement strategies for lowering blood pressure, which included exercise, weight loss, and fish oil. Her medication would be adjusted if the lifestyle changes did not work within three months.
- Recheck blood tests in three months to make sure that fasting blood sugar, HgA1C, vitamin D, C-reactive protein, homocysteine, testosterone, and cholesterol were improving with the supplements and diet.
- If after three months there was no improvement, other treatments would be implemented, including medication to enhance memory. (My bias, after being a psychiatrist for thirty years, is to start with natural treatments, even in serious cases such as Ruth's.)

In talking to Lisa, she told me that seeing her mother this way was a serious wake-up call for her. Lisa's SPECT scan showed mildly low activity in her temporal lobes and parietal lobes (two of the three areas known to be associated with Alzheimer's disease). Her lab testing revealed that she had one of the APOE e4 genes and that her fasting blood sugar was already high as were her BMI and WHtR ratio. To help her mother, and herself, stay on track, she would do the new eating plan with her and make sure her father was totally on board. Support is critical to success.

Three months later, Ruth was doing much better. Her memory test scores had improved as had her weight (she lost 18 pounds and 3 inches off her waist) and all of her important numbers, without any additional medication. At first, getting on the dietary changes was hard for Ruth because she'd never learned how

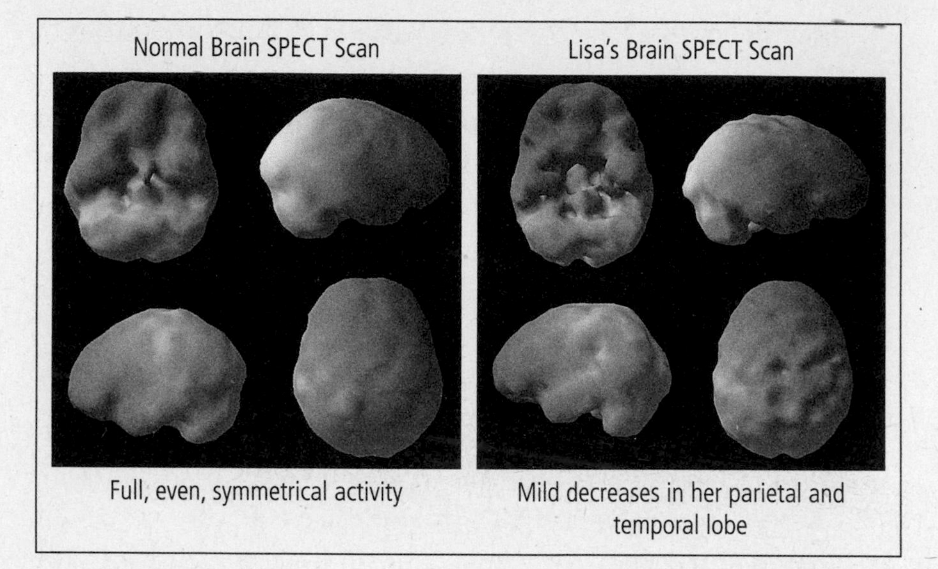

Full, even, symmetrical activity

Mild decreases in her parietal and temporal lobe

to cook in a brain healthy way. Everything had been bread and butter, pancakes and muffins, cakes and cookies. She and Lisa got one of my wife's two cookbooks and used them as a guide. The program kicked in after about two weeks when they realized that healthy food was not only good for them, but that it also tasted great. The program was *not* about being deprived, they learned, but rather learning to prepare an abundance of healthy food. Even on a calorie-restricted diet, their cravings went away and their energy went up. They also loved spending time together, which Lisa realized was more important now than ever before. Given the severity of the diagnosis and seeing Ruth's scans and test scores, Lisa's father also joined the effort and stopped his denial.

To live longer, feel your best, and look and think younger, it is important to do it by the numbers. Knowing your important numbers is a critical step in getting control of your brain and body for the rest of your life. When any or all of these numbers are out of whack, it can prevent you from losing weight, keep you in the dumps, and reduce brain function. Remember the paint-by-number sets when you were little? How you filled in each section, one by one, until a beautiful picture formed? Getting well can be much the same way. The more numbers you get balanced, the more you become a living, breathing, beautiful picture of health. My patients are often surprised to find how fun, rewarding, and motivating it is to see evidence of incremental improvement in black and white on their lab reports or, in the case of SPECT scans, in living color.

DECREASE YOUR RISK FOR ALZHEIMER'S AND OTHER FORMS OF DEMENTIA

I gave Ruth a plan to treat the "early" Alzheimer's disease that was already ravaging her brain, and the treatment had a positive impact for her. I put the word *early* in quotes because in reality the disease in her brain was definitely *not* just starting. Researchers believe that Alzheimer's disease and other forms of dementia actually start decades before people have their first symptoms. In one study from UCLA, brain imaging researchers suggest that brain scans actually start to change as much as fifty years in advance of the illness. The only way we will ever be able to seriously make an impact in decreasing your risk for Alzheimer's disease and other forms of dementia is to start early. Lisa's brain, in her forties, already showed evidence of problems.

Alzheimer's disease is no small problem. It currently affects more than five million people in the United States, and it is estimated to triple by the year 2030. Nearly 50 percent of people who live to the age of eighty-five will develop Alzheimer's. One of the sad truths is that everyone in the family is affected by this disease. The level of emotional, physical, and financial stress in these families is constant and enormous. One of the frightening statistics is that an estimated 15 percent of caregivers of people with Alzheimer's have it themselves.

Decreasing your risk for Alzheimer's disease and other causes of memory loss requires forethought, a well-researched scientific plan (something that will actually work), and a good prefrontal cortex so that you will follow through on the plan. My plan to decrease your risk for Alzheimer's and keep your brain healthy as you age is the whole program in this book. What follows are some obvious steps you can take to decrease the chances your brain will age prematurely.

STEP 1. KNOW YOUR RISK FOR TROUBLE

It is critical to know your specific risk for Alzheimer's disease. Here is a list of the most common risk factors for Alzheimer's disease and early brain aging. The number in parentheses indicates the severity of the risk: the higher the number, the more severe the risk factor. Work to eliminate as many of these risk factors as possible:

One family member with Alzheimer's or dementia (3.5)
More than one family member with Alzheimer's or dementia (7.5)
Family history of Down syndrome (2.7)
A single head injury with loss of consciousness (2.0)
Several head injuries without loss of consciousness (2.0)
Alcohol dependence or drug dependence in past or present (4.4)
Major depression diagnosed by a physician in past or present, whether treated or not (2.0)
Stroke (10)
Heart disease or heart attack (2.5)
High cholesterol (2.1)
High blood pressure (2.3)
Diabetes (3.4)
History of cancer or cancer treatment (3.0)
Seizures in past or present (1.5)
Limited exercise, less than twice a week (2.0)
Less than a high school education (2.0)
Jobs that do not require periodically learning new information (2.0)
Within the 65–74 age range (2.0)
Within the 75–84 age range (7.0)

Over 85 years old (38.0)
Smoking cigarettes for ten years or longer (2.3)
Has one APOE e4 gene, if known (2.5)
Has two APOE e4 genes, if known (5.0)

STEP 2. CONSIDER A SPECT SCAN

Having a brain SPECT scan can also help you know about the health of your brain and your risk of getting Alzheimer's disease. When I turned fifty, my doctor wanted me to have a colonoscopy. I asked him why he didn't want to look at my brain. "Isn't the other end of my body just as important?" How do you really know what is going on in your brain unless you look? At some point in the near future I believe these screening tools will be as normal as mammograms or colonoscopies. They may be especially helpful for people at risk or those having "early" symptoms. Because most physicians are not used to looking at brain function, many people suffer needlessly and never know that they have vulnerable brains until it is too late. They never know that they have a potentially treatable condition.

Here is an example.

Ed

Ed, seventy-two, was brought to us from Vancouver, British Columbia, by his daughter, Candace. Candace was concerned that her father was becoming more forgetful. His mood seemed lower than usual and his judgment was not as good. When she looked at his finances, he had paid bills twice and had forgotten others. When she took him to the local neurologist, the doctor diagnosed him with Alzheimer's disease, without ever looking at his brain. Candace had read my book *Change Your Brain, Change Your Life* and was upset the doctor did not order a scan. The doctor said he was confident in his diagnosis and didn't need to look at his brain. This is always an attitude that amazes me. Unhappy with the lack of thoroughness, Candace brought Ed to see us.

In looking at his scan, Ed had very large ventricles (fluid-filled cavities in the brain). It is a pattern I have labeled the "lobster sign" because what looks like an upside-down lobster appears on the brain slices. Ed also had a small cerebellum in the back, bottom part of the brain. He

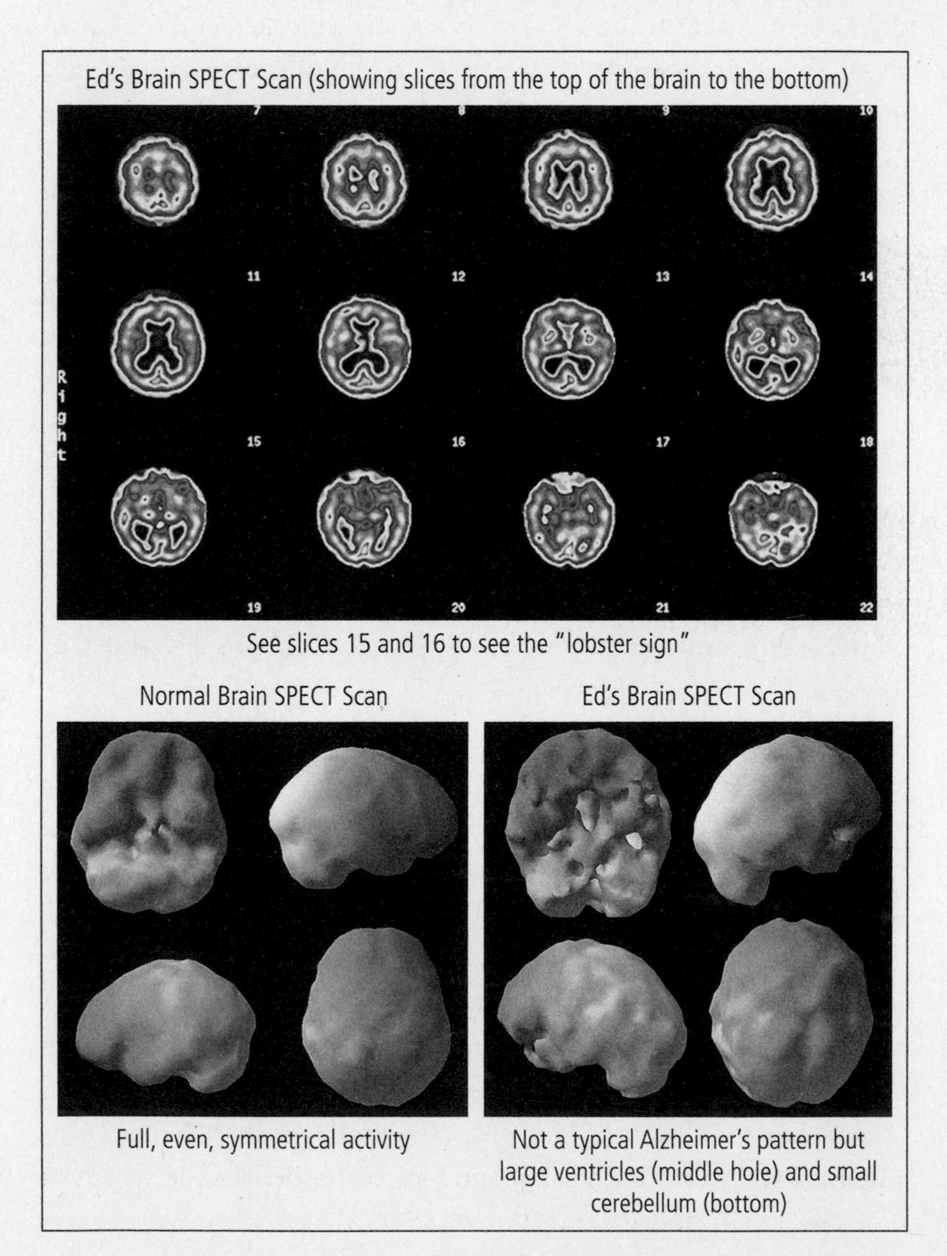

See slices 15 and 16 to see the "lobster sign"

Full, even, symmetrical activity

Not a typical Alzheimer's pattern but large ventricles (middle hole) and small cerebellum (bottom)

definitely did not have the typical Alzheimer's pattern (low temporal and parietal lobe activity).

The reason this finding is so important is that it is often seen with a condition called normal pressure hydrocephalus, or NPH. The drainage of the cerebrospinal fluid is gradually blocked and the excessive fluid builds up slowly over time. It is often associated with urinary incontinence and trouble walking, though not always. Because Ed did not have those other symptoms, his neurologist never thought of NPH. Subsequently, Ed's brain continued to deteriorate. Upon seeing his scan, I recommended he have an immediate neurosurgical consultation. The neurosurgeon agreed with me and placed a shunt in Ed's brain. Within three weeks, Ed's memory came back. How do you know unless you look?

STEP 3. FOCUS ON REDUCING YOUR SPECIFIC RISK FACTORS

Okay, you have an idea of what risk factors you may have. Now what can you do about it? Here is a list of ways to reduce those factors.

Risk: Family member with Alzheimer's disease or related disorder, or you have the APOE e4 gene
Reduce: Seek early screening and take prevention very seriously and as early as possible. Do all of the strategies outlined in this book now.

Risk: Single head injury or several head injuries
Reduce: Prevent further head injuries and start the prevention strategies as soon as possible.

Risk: Alcohol dependence, drug dependence, or smoking in past or present
Reduce: Get treatment to stop and look for underlying causes; start prevention strategies early.

Risk: Major depression or ADHD diagnosed by a physician in past or present
Reduce: Get treatment and start prevention strategies early.

Risk: Stroke, heart disease, high cholesterol, hypertension, diabetes, history of cancer treatment, seizures in past or present
Reduce: Get treatment and start prevention strategies early.

Risk: Limited exercise (less than twice a week or less than thirty minutes per session)
Reduce: Exercise three times a week or more.

Risk: Less than a high school education or job that does not require periodically learning new information
Reduce: Engage in lifelong learning.

Risk: Sleep apnea
Reduce: Seek evaluation and treatment for sleep apnea.

Risk: Estrogen or testosterone deficiency
Reduce: Consider hormone replacement, if appropriate.

You get the idea. Eliminate all the risk factors you can and work to keep your brain healthy over time.

STEP 4. KEEP YOUR BODY AND BRAIN ACTIVE

As we will see in coming chapters, physical and mental exercise is the best way to keep your brain young. Mental exercise helps the brain maintain and make new connections. For more on mental exercise, see chapter 5. Physical exercise boosts blood flow to the brain, improves oxygen supply, helps the brain use glucose more efficiently, and helps protect the brain from molecules that hurt it, such as free radicals. For more on physical exercise, see chapter 3.

STEP 5. SUPPLEMENTS THAT SUPPORT HEALTHY BRAIN FUNCTION

There is a lot of information and misinformation about supplements. Knowing what to do is essential, because some vitamins and supplements work. Take a multivitamin and fish oil every day. Get your vitamin D level checked and optimize it. Here are other supplements that have been shown to be helpful:

- Fish oil and curcumin to decrease inflammation,
- NAC and alpha-lipoic acid to boost the body's ability to handle free radicals and oxidation
- Ginkgo biloba and vinpocetine to boost blood flow
- Huperzine A and acetyl-L-carnitine to boost acetylcholine, the neurotransmitter involved in learning
- Acetyl-L-carnitine and CoQ10 (coenzyme Q10) to boost mitochondrial function (the cells' energy powerhouse, which diminishes with age and aging)
- Brain and Memory Power Boost, a combination of ginkgo, vinpocetine, alpha-lipoic acid, NAC, acetyl-L-carnitine, and huperzine A to boost blood flow and cognitive testing scores when combined with a multiple vitamin, fish oil, and a brain healthy program

STEP 6. EAT TO LIVE LONG

You are what you eat. Many people are not aware of the fact that all of your cells make themselves anew every five months. Food is a drug; intuitively, we all know this. If you have three doughnuts for breakfast, how do you feel thirty minutes later? Blah! If you have a large plate of pasta for lunch, how do you feel at 2 p.m.? Blah! The right diet helps you feel good. The wrong diet makes you feel bad. Diet is an extremely important strategy to keep your brain healthy with age.

The best diet is one that is high in nutrients, low in calories (calorie restriction is associated with longevity), high in omega-3 fatty acids (fish, fish oil, walnuts, and avocados), and antioxidants (vegetables). The best antioxidant fruits and vegetables, according to the U.S. Department of Agriculture, include prunes, raisins, blueberries, blackberries, cranberries, strawberries, spinach, raspberries, Brussels sprouts, plums, broccoli, beets, avocados, oranges, red grapes, red bell peppers, cherries, and kiwis. Eat your fruits and vegetables! Your mother was right. For much more on nutrition, see chapter 2.

If you follow these guidelines you can decrease your risk for Alzheimer's disease by 50 percent or more. In a new study from San Francisco researchers found that if people exercised and eliminated smoking, hypertension, depression, and obesity, they could dramatically cut their chances of getting Alzheimer's disease and other problems of aging. It's your choice. Do you want a good mind going forward or not? If you do, now is the time to get serious, not at some arbitrary time in the future.

SIMPLE STEPS TO DRAMATICALLY DECREASE YOUR RISK OF ALZHEIMER'S

Of course, with any recommendation, especially one with such a high promise as decreasing your risk for Alzheimer's disease, it is important to ask the question "How do you know?" The strategy of decreasing your risk factors to decrease your chances of getting Alzheimer's was studied by Dr. Deborah Barnes, a mental health researcher at the San Francisco VA Medical Center. Her research found that *more than half of all Alzheimer's disease cases could potentially be prevented through lifestyle changes and treatment, or prevention of chronic medical conditions.* She looked at research on hundreds of thousands of Alzheimer's disease patients from around the world and concluded that many risk factors for Alzheimer's disease can be reduced.

Worldwide, the risk factors for Alzheimer's disease include low education, smoking, physical inactivity, depression, midlife hypertension, diabetes, and midlife obesity. In the United States, Dr. Barnes found that the biggest risk factors that are modifiable are physical inactivity, depression, smoking, midlife hypertension, midlife obesity, low education, and diabetes. Her research concluded that this group of risk factors is believed to account for up to 51 percent of all Alzheimer's patients worldwide (17.2 million cases) and up to 54 percent of Alzheimer's cases in the United States (2.9 million cases). Her research suggests that some very simple lifestyle changes can have a dramatic effect on preventing Alzheimer's and other dementias in the United States and worldwide. The study results were presented at the 2011 meeting of the Alzheimer's Association International Conference on Alzheimer's Disease in Paris and published online in *Lancet Neurology*.

REAL PREVENTION STARTS WITH OUR CHILDREN

Many of the risks for Alzheimer's disease occur in childhood. If we are sincere about preventing this disease and related problems we must start with our children. The APOE e4 gene increases the risk of Alzheimer's. Having this gene plus a head injury increases the risk further. Many head injuries occur in childhood, especially when playing contact sports or doing other high-risk activities. If children are allowed to engage in these activities they should first be screened for the APOE e4 gene. If they have it, we should be more attentive to protecting their heads. Children with ADHD and learning problems often drop out of school, leaving them at higher risk for dementia. Making sure we properly diagnose and help these children is essential to helping them become lifelong learners. The seeds for depression occur in childhood. Depression is often a result of persistent negative thinking patterns.

School programs should be developed to teach children how to correct these patterns, which could help decrease depression. Childhood obesity leads to adult obesity. Educating children on nutrition and exercise can have lifelong benefits.

Get healthy for yourself, but also do it for your children and great-grandchildren who want you around, lucid and vibrant and smiling, to enjoy life with them as long as humanly possible.

CHANGE YOUR AGE NOW: TWENTY BRAIN HEALTHY TIPS AND NUMBERS TO KNOW FOR A LONG, HEALTHY LIFE

1. **Know your important health numbers.** Get tested and write them down, make lifestyle changes to improve them, and retest every three months until you are out of any danger zones. Some of the numbers you'll want to know are listed below.

2. **Know your BMI.** Lose weight if you are overweight by eating low-calorie but highly nutritious foods. Think CROND: calorie-restricted, optimally nutritious, and delicious. Being obese has been associated with less brain tissue and lower brain activity and doubles the risk for Alzheimer's disease. Make sure your waist size is half your height in inches. Yes, you actually have to measure your waist with a tape measure.

3. **Get your five to ten fruits and veggies per day.** Count them! Eat more vegetables than fruits and try to get that number to between five and ten servings to enhance your brain and lower your risk for cancer.

4. **Get eight hours' sleep every night.** Getting fewer than eight hours has been associated with cognitive decline. Chronic insomnia triples your risk of death from all causes. Try a warm bath before bed, no television an hour before bed, a sleep-inducing hypnosis CD,

and a melatonin-based sleep supplement. Sleep apnea doubles a person's risk for Alzheimer's disease, so get a sleep study done if you suspect this could be an issue.

5. **Check your blood pressure often, and make sure it is under control.** If it is high, fish oil and exercise, along with losing weight, can help. If this protocol doesn't help, make sure to see your health care professional. High blood pressure is an emergency and is the second-leading preventable cause of death in the United States, after smoking.

6. **If you smoke, quit.** It prematurely ages the brain and body.

7. **I am not a fan of alcohol intake, because of what I see on brain scans.** Don't overdo it.

8. **Get a complete blood count.** Low blood count can make you feel anxious and tired and can affect your memory. One of our patients got this as a screening test and found out he had leukemia. Early treatment works better than late treatment.

9. **Get a general metabolic panel.** This will test the health of your liver, kidneys, fasting blood sugar, and cholesterol. Every organ in your body is related to brain health and vice versa.

10. **Get an HgA1c test.** It shows your average blood sugar levels over the past two to three months and is used to diagnose diabetes and prediabetes. Alpha-lipoic acid has been shown to be helpful to stabilize blood sugar.

11. **Check your 25-hydroxy vitamin D level. This is critical and easy to fix. Also, check your folic acid and B_{12} levels.** A deficiency in these vitamins can add to cognitive decline.

12. **Know your thyroid levels.** Abnormal thyroid hormone levels are a common cause of forgetfulness, confusion, lethargy, and other symptoms of dementia in both women and men. Having low thyroid levels decreases overall brain activity.

13. **Find out your C-reactive protein level.** This is a measure of inflammation. Elevated inflammation is associated with a number of diseases and conditions that are associated with aging and cognitive impairment. The eating lifestyle I encourage in this book is also an anti-inflammatory diet. Fish oil also helps lower inflammation.

14. **Find out your homocysteine levels.** Elevated levels in the blood have been associated with poor arterial health, as well as a possible increased risk of heart attacks, strokes, blood clot formation, and possibly Alzheimer's disease. Lentils, asparagus, spinach, and most beans are good for lowering your homocysteine levels, along with folic acid (1 mg), B_6 (10 mg), and B_{12} (500 μg).

15. **Test for excess ferritin.** This is a measure of excess iron stores that increases with inflammation and insulin resistance. Higher iron stores have been associated with stiffer blood vessels and vascular disease.

16. **Know your free and total serum testosterone level.** Low levels of the hormone testosterone, for men or women, have been associated with low energy, cardiovascular disease, obesity, low libido, depression, and Alzheimer's disease.

17. **Don't ignore or minimize increased forgetfulness or downplay it with "She's just ditzy" or "He's just getting older."** Forgetfulness and fogginess could have a number of causes (from adult

attention deficit disorder, or ADD, to anemia, to early-onset dementia)—but they are signs, at minimum, that your brain can use a tune-up.

18. **Technology, like GPS systems, may be delaying the diagnosis of dementia because it can conceal forgetfulness.** Make sure you get regular memory screenings after the age of fifty. You can do this at www.theamensolution.com.

19. **Dramatically decrease your risk of Alzheimer's disease by decreasing all of the risk factors that are associated with it.** These include diabetes, heart disease, obesity, depression, brain trauma, and cancer.

20. **Real prevention starts with getting our children healthy.**

2

TAMARA

FOCUS ON FOODS THAT SERVE YOU, NOT STEAL FROM YOU

The major cause of inflammation is our processed high-sugar, low-fiber, fast-food, junk-food, calorie-dense, nutrient-poor industrial diet and our couch-potato lifestyle. A plant-based, whole-foods, real-food diet without sugar and flour in pharmacologic doses along with anti-inflammatory omega-3 fats and a good dose of exercise can dramatically reduce the risk of and even reverse heart disease and diabetes. And they cost a lot less.

—MARK HYMAN, M.D.

Food is medicine or it is poison. You get to choose.

Research suggests that your diet will affect generations to come. In a new study, animals fed the typical Western diet, over several generations, bred offspring who were born more obese. The diet actually changed the way the animal's genes functioned so that over time each generation ate more food and became fatter and fatter. This is very disturbing, because it means that unless you get your health under control now, it can affect your children, your grandchildren, and even your great-grandchildren.

The opposite is also true. As you care for yourself, you are increasing the odds that your children and grandchildren will have better health. Scientists, researchers, and psychiatrists are finding that everything we do has ripple effects on the emotional and physical health of our families and beyond. You have the ability to change your life and the lives of generations to come.

AN OLD WOMAN AT AGE THIRTY-TWO

For Tamara, my wife's sister, it was a miserable day that began like others from the previous year. At the time, Tamara was a thirty-two-year-old mother of a five-year-old and a baby. She was unable to sleep through the night and would wake at intervals with a fierce craving for sugar. On more than one middle-of-the night excursion to the kitchen, unable to find something to satisfy her raging sweet tooth, Tamara ate the kids' sweet-tasting vitamins.

Her numerous trips to the toilet would begin in earnest around 4:00 a.m. every day. The contents of her bowels were the consistency of water, and at last count, the day before this one, she'd visited the bathroom thirteen times. She was spending three to five hours a day on the toilet. Every joint in her body ached like the body of an old arthritic woman, and with every movement, her muscles burned as if on fire. Her hands were so swollen that as she changed her baby's diaper that morning, she saw that her knuckles were beginning to crack and bleed—her skin unable to contain whatever toxic monster within was blowing her body up like a balloon. Her stomach was so painfully bloated most of the time that she still looked pregnant.

The complete lack of energy made working impossible. Taking her young kids to the park, something she longed to do on pretty days, was unthinkable. It was all she could do to care for the children's most basic needs as she lay on the couch most of the day, with zero energy and terrible pain soaring with every movement. "There wasn't an energy drink or a strong enough brew of coffee that would even faze me anymore,"

Tamara said. Her husband, Hector, was a wonderful man, compassionate and hardworking. He was also terribly concerned about what was happening to his wife. Many of the older people Tamara loved and had cared for in her family had recently died, and the toll of these multiple losses, on top of her waning health, hung over her like a heavy cloud. She'd begun taking an antidepressant, hoping it might help dispel some of the darkness that shrouded her days and nights.

For whatever reason, this particular day in November of 2010, Tamara hit a wall. She'd finally had enough. She realized if she didn't do something radical, she could die; and her husband would lose the love of his life, and her children would grow up without knowing their mother. As she lay on the couch, tears streaming, she thought of her big sister, Tana—my wife. Tamara knew that Tana was a fitness and nutrition coach, the picture of health. But she also remembered that Tana had battled serious health issues in her past and had found a way to overcome them. Tana had never pushed her nutritional or exercise philosophy on her extended family, but Tamara had always known in her heart that Tana would be happy to help if and when she was ready to get serious about health. Tamara knew it was time to ask for help; she reached for her cell phone and made the call that would soon change her life.

TIRED OF PILLS AND LABELS

Shortly before this life-changing phone call, Tamara saw a doctor for her myriad issues. She'd been suffering with an unexplained autoimmune disorder for years. At one time doctors suspected she had lupus. When that was ruled out, they tested her for rheumatoid arthritis and then several other diseases. Eventually, she fell into the catchall diagnosis of fibromyalgia. But she didn't know many people with fibromyalgia who spent several hours every day in the bathroom. This doctor put her through another battery of tests to rule out rheumatoid arthritis (again), celiac-sprue, Crohn's disease, and irritable bowel syndrome. Lab tests showed her triglycerides were 290 mg/dL (normal is less than 150 mg/dL), cholesterol was 250 mg/dL (less than 200 mg/dL is optimal), and her blood pressure was 139/96 mm/Hg (below 120/80 mm/Hg is optimal)! At 5'4", her weight had crept up to 206 pounds. Her vitamin D level was less than 20 ng/mL (optimal is 50–90 ng/mL). Even with these numbers in hand, the doctor said nothing to Tamara about a lifestyle change but was quick with the prescription pad, ordering medications for her array of symptoms.

As sick as Tamara was, she left the doctor's office with foreboding. Something in her hesitated to take the prescribed medications. In her experience, most of the medical world had given her either labels or pills, and the outcome of both had never been good. She remembered all too well the nightmare of getting off pain pills, which had been prescribed for her in the past, and knew she could not go there again.

I WANT YOU TO EAT LIKE A GORILLA

Tana's voice was now on the other end of the phone, and she listened compassionately as Tamara shared her despair. Tana understood her sister's pain and frustration, as she too had suffered from a variety of health issues before discovering she had several food allergies. She suspected that since she and Tamara were half sisters and shared genes, Tamara might also have severe and undiagnosed food allergies. At this point Tamara was all ears, ready to try anything.

"It is a choice," Tana said. "If you are ready to make the decision to radically improve your health, I want you to begin an elimination diet along with a few supplements. And Tamara, I want you to start today." Tana then asked her sister what sort of food she and her family had been eating.

"Oh, the basic American diet," Tamara quipped. "Whatever comes out of a box."

"Do you eat any vegetables?"

"Honestly, Tana, I can't remember the last time I ate a vegetable."

Tana took a deep breath. "Okay, get ready for a big change. I want you to start eating like a gorilla."

Tamara laughed, but Tana was on a roll. One thing about these talkative sisters, they are never at a loss for words and when they have something to say, they say it with passion, and they talk fast. Tana said, "I want you to eat all the green vegetables you can manage to eat, lightly cooked or in salads. I want you to eat three palm-sized portions of protein a day. I also want you to drink ten to twelve glasses of water a day. Two glasses before you eat breakfast. You need it to wash those toxins out of your body. Your body is full of inflammation and it is killing you. This is the fastest way to get rid of it. You can also throw in some nuts, seeds, and eventually berries when you start to add fruit into your routine. But for now, avoid all fruit, until we get that raging sugar craving of yours completely under control. Also, no dairy."

Tamara sighed. "I'm going to have to empty my kitchen cabinets. They are full of junk, and I know if there's any sugar in the house, I'm helpless to avoid it right now."

Tana agreed this was a good idea and encouraged her sister to "make her fridge into a rainbow," so Tamara tossed out the chips and bought some hummus and a colorful array of bell peppers. Because their family was on a tight budget it took some shopping around to find affordable, healthy food. Tamara hadn't bought vegetables in such a long time that she hadn't realized how expensive they were. But she found organic spinach in bulk at a popular discount warehouse and ground turkey was lean, tasty, and affordable. For the first two days, she fought through the cravings, dining on her "gorilla zoo food" of fresh green salads, accompanied by turkey patties with a creamy slice of avocado on the side. She loved cheese, so that was hard to give up, but she'd never liked milk and found the taste of almond and coconut milk to be delightful. For everything she gave up, she discovered a new and more nutritious food that she really liked, a food that satisfied her, made her feel better, and didn't leave her with irrational powerful cravings for more.

Tana also sent Tamara some nutritional powdered shake mixes and some supplements, including a high-quality fish oil and vitamin D from the Amen Clinic. She also sent our specially formulated supplements designed to help with cravings and sharpen her focus and energy without the unwanted side effects of caffeine.

THE TWO-DAY MIRACLE

Two days later, a miracle happened. It felt like a miracle to Tamara anyway. She went to the bathroom and afterward called Tana, in tears. "Tana! I had my first normal bowel movement in months! I can't believe it! Can this be a result of just a change in diet? Surely this is a fluke. I know it is crazy to be this happy about 'normal poop'—but I've been living such a nightmare with this embarrassing, exhausting, chronic diarrhea for so long, I just can't tell you how great this is to me!"

Tana laughed and encouraged her sister, telling her, "Don't worry, we're not nearly done yet. This is only the beginning." Within two months, Tamara had also lost seventeen pounds. The goal had not been to lose weight per se. The goal was to save Tamara's health and possibly her life. But the by-product of weight loss was so much fun for Tamara, and her self-esteem soared along with her renewed energy. "I can suck in my stomach now!" she says. "Before, I was always so bloated, I couldn't even feel or find the muscles to contract them."

Tamara returned to her doctor, thrilled to tell her about her progress, assuming the physician would give her high fives and ask what she'd done to turn her health around. To Tamara's shock, it was as if the doctor didn't hear anything she said. In fact the doctor said, "You are young and you need this medication. Diet alone won't cure your issues. You will probably need to be on medication for the rest of your life."

Before leaving the office, Tamara asked for a copy of her latest blood work and left, still stunned about the absence of encouragement from her physician.

She called Tana and shared what had happened, explaining her reluctance to go on medication when diet changes alone had accomplished so much, so fast. Then Tana asked about the blood work. Tamara faxed it over, and within minutes Tana was on the phone, her voice happy and excited. "Tamara, this is unbelievably good news. Your triglycerides have gone from 295 to 129. Your cholesterol has gone from 258 to 201. Your blood pressure is now normal, dropped from 136/94. I can't believe your doctor didn't share and rejoice in this improvement with you!"

Tana continued, "I'm not telling you not to take the medicine. But let's experiment by continuing what you are doing, without meds, for another month. Then let's revisit this."

THE FUTURE'S SO BRIGHT, SHE'S GONNA NEED SHADES

Within a month, Tamara had improved so much she was like a new person. The medications still remain unopened, unused. Twenty pounds lighter now, she has plenty of energy. When her husband was laid off recently from his job due to economic downturn, Tamara waited for the stress of that to sink her, but to her surprise it didn't. She was coping well. So well, in fact, she was able to get a job as a waitress at a nice family restaurant—a job requiring lifting heavy trays and staying on her feet—to help with emergency income. In addition, she is also regularly walking around a lake near her house with some of her girlfriends from work, many of whom struggle with fibromyalgia-like symptoms and weight issues too. "I don't preach to them," says Tamara, "because I'm still learning too, but I know that watching my life has inspired them to make changes in their health as well. And when they ask me, I'm ready with answers."

Hector is a fabulous cook and his specialties are spicy, flavorful Mexican dishes from his Hispanic heritage. Now he uses many of the same spices and flavors of the Southwest, but adds lots of veggies to his dishes and makes large fresh salads to go with them. The children easily adapted to a healthier way of eating and especially love fresh fruit. As of this writing, Tamara looks and feels dramatically younger than she did just seven months earlier. And not only Tamara: Her new behavior has rubbed off on her children, and if she continues to model brain healthy nutrition she will have a positive effect on her grandchildren as well.

FOOD, LONGEVITY, BIZARRE TIMES, AND THE LITTLE LIES

If you want to live longer, look younger, and be healthier, happier, and smarter, getting your food under control is critical. In fact, it may be *the* most important factor in increasing your longevity odds. Food is medicine, or as we saw in Tamara's case, it can be poison. The changes in her diet were critical to her success and if she sticks with it for the long haul it will likely save and extend her life.

Getting your food under control in our current society is no small feat, as we live in a very bizarre time. Nearly everywhere we go we are being pounded by the wrong messages about food that will make us fat, depressed, and feebleminded. Look what we're bombarded with:

- Foot-long hot dogs at the ball park
- Huge meal portions at restaurants
- Food pushers who ask us to supersize everything for less money
- Monster foods pushed by billboards

I was recently driving down the 405 freeway in Los Angeles when I saw a billboard for a huge fast-food sandwich.

And, then, no lie, as I turned my head to the other side of the freeway I saw another billboard for losing weight with Lap-Band.

Indulge yourself, continually make bad decisions, and then pretty soon you or your patients will need surgery to get your inner child under control. Sounds pretty crazy. We need a better way, which is to start being smart about our nutrition and stop the lies that perpetuate illness.

Here are some of the recent lies I have heard people tell themselves about why they have to eat poor-quality food, and my response in parentheses:

- "I can't eat healthy because I travel." (I am always amused by this one, as I travel a lot for my public television shows. It just takes a little forethought and judgment when ordering.)
- "I work hard and eat out a lot." (Probably not harder than me or many other really healthy people.)
- "My whole family is fat; it is in my genes." (This is one of the biggest lies. It is our behavior that triggers the expression of our genes. My genes too say I should be fat, but I do not give in to the behavior that would make me so.)
- "My family won't cooperate." (If your family started to do cocaine or steal, would you join them? It is better if you eat healthy together, but ultimately you are responsible for you.)
- "It's my boss's fault." (For the ineffective person, it is often someone else's fault. When you take responsibility to be better, you are much more likely to win.)
- "It's Easter, Memorial Day, the Fourth of July, Labor Day, Thanksgiving, Christmas, New Year's Day, my birthday, my dog's birthday. It is Monday, Tuesday, Wednesday, Thursday, Friday, Saturday, or Sunday." (There is always a reason to cheat, to celebrate, or to commiserate. You get healthy with food when you realize it is medicine or poison.)
- "I'll start tomorrow!" (I used to say this year after year, then realized tomorrow never comes.)
- "My son will only eat Cap'n Crunch for breakfast." (I recently did an exercise with my wife in which together we cleaned out the kitchen of a woman at our church. She said she was mostly a vegetarian and ate a healthy diet. Her cabinets were loaded with bad foods. When we got to the Cap'n Crunch she told us that was

the only food her teenage son would eat for breakfast. "Really?" I asked. "If you did not have Cap'n Crunch, he would starve?" At that point, she turned her head away and said, "No. He would find something else that was healthy." The little lies do not only hurt us, they hurt the people we love.)

- "Good food is expensive." (Being sick is really expensive. In fact, the cheapest food is the most expensive in terms of illness and lost productivity.)
- "I would rather get Alzheimer's disease, heart disease, cancer, or diabetes than give up sugar." (This one always shocks me. But I understand addiction. Certain combinations of fat, sugar, and salt actually work on the heroin centers of the brain and can be totally addictive.)

You literally are what you eat. Throughout your lifetime, your body is continually making and renewing its cells, even brain cells. Your skin cells make themselves new every thirty days! So eating right can have a dramatic, positive impact on your skin in a short period of time. Food fuels cell growth and regeneration, so what you consume on a daily basis directly affects the health of your brain and body.

In addition, you likely have already noticed how foods affect your mood and energy level. Ever feel buzzed for an hour, then ready for a long winter's nap after a Grande Americano? Have you ever felt shaky, foggy, headachy, or weak after a breakfast of orange juice with pancakes drowned in syrup? How about that stuffed feeling after you've eaten too much of anything, and just want to lie down and sleep it off? Too much white bread and protein, and an absence of fresh fruit and vegetables, is constipation waiting to happen, which is one of the reason people struggle with a sluggish digestion when they travel. (Lack of exercise is another.) They're grabbing burgers on white bread or chicken nuggets and downing them with a soft drink loaded with high-fructose corn syrup (or a yeasty beer) as they pick up their dinner at the drive-through or an airport kiosk. In short, if you want to feel old, weak, bloated, cranky, sleepy, and constipated, this is the diet for you!

CHEAT DAYS NOT REQUIRED OR ENCOURAGED

If you want to feel and look happier, healthier, and younger in the future, be a warrior for the right nutrition. Cheating and cheat days are not required or encouraged. I am often amazed at the number of health programs that encourage cheat days. Since food can clearly be addictive, would it ever make sense to tell any addict that it is okay to have cheat days? Imagine telling a cocaine addict or smoker or alcoholic to have cheat days. That could trigger a relapse. What about a sex addict? Would it be okay for her to take a day off and grab a few prostitutes? Will that help reset her metabolism? Use your brain, and good common sense, to change your age.

SEVEN RULES FOR BRAIN HEALTHY EATING FOR THE LONG RUN

Over the years I have been able to distill our message about brain healthy nutrition into seven simple rules. If you follow them, food will become your longevity medicine.

1. Eat high-quality calories and not too many of them.
2. Drink plenty of water and avoid liquid calories.
3. Eat high-quality lean protein.
4. Eat "smart" (low-glycemic, high-fiber) carbohydrates.
5. Limit fat consumption to healthy fats, especially those containing omega-3s.
6. Eat natural foods of many different colors to boost antioxidants.
7. Cook with brain healthy herbs and spices.

RULE 1. EAT HIGH-QUALITY CALORIES AND NOT TOO MANY OF THEM

My wife will tell you that I am not cheap, but I always want great value for the money I spend. I think of calories the same way. I always want high-quality nutrition that serves my mind and body. You should too. I want you to think of eating and drinking only high-quality calories and not too many of them. There is extensive research on this concept for longevity. Restricting calories not only helps you control weight, but it decreases the risk for heart disease, cancer, diabetes, and stroke. Even better, restricting calories triggers certain mechanisms in the body to increase the production of nerve growth factors, which are beneficial to the brain. To get the most out of your food, think CROND (calorie restricted, optimally nutritious, and delicious). This means that once you figure out how many calories are optimal for you to maintain your weight (or lose weight if needed), each of those calories needs to be jam-packed with nutrition. The other benefit of eating nutrition-laden foods in moderation is that they are naturally more filling. So don't let calorie restriction make you think "starving" or even "hungry"—because if you eat well, the fiber and protein and clean fresh water will keep you plenty satiated; and you'll also feel light, lean, and energetic.

Another one of the little lies professionals will tell you is that you do not need to count your calories. Of course you do if you want to stay healthy. Not counting your calories is like not knowing how much money you have in the bank while you continue to spend, spend, spend until your body becomes bankrupt. By knowing what you are putting into your body and using the CROND principle, you are much more likely to be healthy and live longer.

The average fifty-year-old woman needs about 1,800 calories a day to maintain her weight, and the average fifty-year-old man needs 2,200 calories a day. Of course, it will vary depending on height and activity level. Go to www.amenclinics.com for a free calorie calculator.

Attitude here is critical. If you become a calorie value spender you can eat great food and feel more than satisfied and avoid foods that hurt you. The typical Western diet of bad fat, salt, and sugar (think cheeseburgers, fries, and sodas) promotes inflammation and has been associated by itself with depression, ADD, dementia, heart disease, cancer, diabetes, and obesity. But if you start making better choices today you will quickly notice that you have more energy, better focus, a better memory, better moods, and a slimmer, sexier waistline. A number of new studies have reported that a healthy diet is associated with dramatically lower risks of Alzheimer's disease and depression. What really surprised me when I decided to get healthy, and really learned about food, my food choices got better, not worse.

It was the start of a wonderful relationship with food. I was no longer a slave to foods that were hurting me. I used to be like a yo-yo. Crave bad food . . . overeat it . . . feel lousy . . . then hate myself in the process. It was way too much drama. Since I have been on my program, I have never eaten better and it affects everything in my life in a positive way. My wife, Tana, has written several amazing cookbooks to help people get thinner, smarter, and happier, and of course I get to try everything first. I love her lentil soup and stuffed bell peppers, think of berry nutty quinoa as dessert, and feel smarter with fresh wild salmon. I don't want fast food anymore because it makes me tired and stupid. Now I want the right food that makes me smarter. And, contrary to what most people think, eating in a brain healthy way is not more expensive—it is less expensive. My medical bills are lower and my productivity has gone way up. And what price can you put on feeling amazing? Be smart. Use food as medicine that heals you.

RULE 2. DRINK PLENTY OF WATER AND AVOID LIQUID CALORIES

Your brain is 80 percent water. Anything that dehydrates it, such as too much caffeine or alcohol, decreases your thinking and impairs your judgment. Make sure you get plenty of filtered water every day.

On a trip to New York City recently I saw a poster that read "Are You Pouring on the Pounds . . . Don't Drink Yourself Fat." I thought it was brilliant. A recent study found that, on average, Americans drink 450 calories a day, twice as many as we did thirty years ago. Just adding the extra 225 calories a day will put 23 pounds of fat a year to your body, and most people tend *not* to count the calories they drink. Did you know that some coffee drinks or some cocktails, such as margaritas, can cost you more than 700 calories? One very simple strategy to get your calories under control to live longer is to eliminate most of the calories you drink. My favorite drink is water mixed with a little lemon juice and a little bit of the natural sweetener stevia. It tastes like lemonade, so I feel like I'm spoiling myself, and it has virtually no calories.

Proper hydration is a very important rule of good nutrition. Even slight dehydration increases the body's stress hormones. When this happens, you get irritable, and you don't think as well. Over time, increased levels of stress hormones are associated with memory problems and obesity. Dehydration also makes your skin look older and more wrinkled. Water also helps to cleanse the body of impurities and toxins.

Make sure your water is clean. Having a water filter on your drinking water faucets at home and only drinking from phthalate and BPA-free water bottles is best.

Be aware that not all liquids are created equal. It is best to drink liquids that are low in calories or have none at all and that are free of artificial sweeteners, sugar, caffeine, and alcohol. I also encourage my patients to drink green tea (unsweetened or lightly sweetened with stevia) two or three times a day. Caffeinated green tea has half the calories of coffee, so it is not terrible. Decaffeinated green tea is an alternative if you are eliminating caffeine from your diet (which is a good choice for many people). Researchers from China found that when people drank two to three cups of green tea a day, their DNA actually looked younger than that of those who did not.

RULE 3. EAT HIGH-QUALITY LEAN PROTEIN

Protein helps balance your blood sugar and provides the necessary building blocks for brain health. Protein contains L-tyrosine, an amino acid that is important in the synthesis of brain neurotransmitters. Found in foods like meat, poultry, fish, and tofu, it is the precursor to dopamine, epinephrine, and norepinephrine, which are critical for balancing mood and energy. It is also helpful in the process of producing thyroid hormones, which are important in metabolism and energy production. Tyrosine supplementation has been shown to improve cognitive performance under periods of stress and fatigue. Stress tends to deplete the neurotransmitter norepinephrine, and tyrosine is the amino acid building block to replenish it.

Also found in protein is L-tryptophan, an amino acid building block for serotonin. L-tryptophan is found in meat, eggs, and milk. Increasing intake of L-tryptophan is very helpful for some people in stabilizing mood, improving mental clarity and sleep, and decreasing aggressiveness.

Eating protein-rich foods like fish, chicken, and beef also provides the amino acid glutamine, which serves as the precursor to the neurotransmitter GABA (gamma-aminobutyric acid). GABA is reported in the herbal literature to work in much the same way as antianxiety drugs and anticonvulsants. It helps stabilize nerve cells by decreasing their tendency to fire erratically or excessively. This means it has a calming effect for people who struggle with temper, irritability, and anxiety.

Great sources of lean protein include fish, skinless turkey, chicken, and lean beef (hormone free, antibiotic free, free range), beans, raw nuts, high-protein grains, and high-protein vegetables, such as broccoli and spinach. Did you know that spinach is nearly 50 percent protein? I use it instead of lettuce on my sandwiches for a huge nutrition boost.

It is especially important to eat protein at breakfast because it increases attention and focus, which we need for work or school.

Eating carbohydrates boosts serotonin in the brain, which induces relaxation, and that makes you want to sleep through your morning meetings. In the United States, we have it backward. We tend to eat high-carbohydrate cereal, pancakes, or bagels for breakfast and a big steak for dinner. Doing the opposite may be a smarter move for your brain. I love the idea of using food to fuel your ability to focus or relax. If I need to work at night I will increase my protein. If it has been a stressful day, I am more likely to eat a higher concentration of carbohydrates to calm my brain.

RULE 4. EAT "SMART" (LOW-GLYCEMIC, HIGH-FIBER) CARBOHYDRATES

Eat carbohydrates that do not spike your blood sugar and are high in fiber, such as whole grains, vegetables, and fruits like blueberries and apples. Carbohydrates are not the enemy. They are essential to your life. Bad carbohydrates are the enemy. These are carbohydrates that have been robbed of any nutritional value, such as simple sugars and refined carbohydrates.

Get to know the glycemic index (GI). The glycemic index rates carbohydrates according to their effects on blood sugar. It is ranked on a scale from 1 to 100-plus with the low-glycemic foods having a low number (which means they do not spike your blood sugar, so they are generally healthier for you) and the high-glycemic foods having a high number (which means they quickly elevate your blood sugar, so they are generally not as healthy for you).

Eating a diet that is filled with low-glycemic foods will lower your blood glucose levels and decrease cravings. The important concept to remember is that high blood sugar is bad for your brain, and ultimately your longevity.

Be careful when going by GI to choose your foods. Some low-glycemic foods aren't good for you. For example, in the following list, you might notice that peanut M&Ms have a GI of 33 while steel-cut oatmeal has a GI of about 52. Does this mean that it's better for you to eat peanut M&Ms? No! Peanut M&Ms are loaded with sugar, saturated fat, artificial food coloring, and other things that are not brain healthy. Steel-cut oatmeal is a high-fiber food that helps regulate your blood sugar for hours. Use your brain when choosing your food.

In general, vegetables, fruits, legumes, and nuts are the best low-GI options. A diet rich in low-GI foods not only helps you lose weight, it has also been found to help control diabetes, according to a 2010 review of the scientific literature in the *British Journal of Nutrition*. Be aware, however, that some foods that sound healthy actually have a high GI. For example, some fruits like watermelon and pineapple have a high ranking. It is wise to consume more fruits on the low end of the spectrum. Similarly, some starches like potatoes and rice, and some high-fiber products like whole wheat bread, are on the high end of the list. Eating smaller portions of these foods and combining them with lean proteins and healthy fats can reduce their impact on blood sugar levels.

The following list of foods and their GI is culled from numerous sources, including a 2008 review of nearly twenty-five hundred individual food items by researchers at the Institute of Obesity, Nutrition and Exercise in Sydney, Australia. Make a copy of this list and keep it with you when you go grocery shopping.

GLYCEMIC INDEX (GI)

Low GI	55 and under
Medium GI	56–69
High GI	70 and above

GLYCEMIC INDEX RATINGS

Grains	**Glycemic Index**
White bread	75 ± 2
Whole wheat bread	74 ± 2
White rice	72 ± 8
Bagel, white	69

Brown rice	66 ± 5
Couscous	65 ± 4
Basmati rice	57 ± 4
Quinoa	53
Pumpernickel bread	41
Barley, pearled	25 ± 2

Breakfast Foods	**Glycemic Index**
Scones	92 ± 8
Instant oatmeal	79 ± 3
Cornflakes	77
Waffles	76
Froot Loops	69 ± 9
Pancakes	66 ± 9
Kashi Seven Whole Grain Puffs	65 ± 10
Bran muffin	60
Blueberry muffin	59
Steel-cut oatmeal	52 ± 4
Kellogg's All-Bran	38

Fruit *(raw unless otherwise noted)*	**Glycemic Index**
Dates, dried	103 ± 21
Watermelon	80 ± 3
Pineapple	66 ± 7
Cantaloupe	65
Raisins	64 ± 11
Kiwi	58 ± 7
Mango	51 ± 5
Banana, overripe	48
Grapes	43
Nectarines	43 ± 6
Banana, underripe	42
Oranges	45 ± 4
Blueberries	40

Strawberries	40 ± 7
Plums	39
Pears	38 ± 2
Apples	36 ± 5
Apricots	34 ± 3
Peach	28
Grapefruit	25
Cherries	22

Vegetables	**Glycemic Index**
Baked potato	86 ± 6
Sweet potato	70 ± 6
Sweet corn	52 ± 5
Peas	51 ± 6
Carrots, boiled	39 ± 4
Artichoke	15
Asparagus	15
Broccoli	15
Cauliflower	15
Celery	15
Cucumber	15
Eggplant	15
Green beans	15
Lettuce	15
Peppers	15
Snow peas	15
Spinach	15
Squash	15
Tomatoes	15
Zucchini	15

Legumes and Nuts	**Glycemic Index**
Baked beans, canned	40 ± 3
Chickpeas	36 ± 5
Pinto beans	33
Butter beans	32 ± 3

Lentils	29 ± 3
Cashews	25 ± 1
Mixed nuts	24 ± 10
Kidney beans	22 ± 3

Beverages	**Glycemic Index**
Gatorade orange flavor	89 ± 12
Rice milk	79 ± 8
Coca-Cola	63
Cranberry juice	59
Orange juice	50 ± 2
Soy milk	44 ± 5
Apple juice, unsweetened	41
Milk, full fat	41 ± 2
Skim milk	32
Tomato juice	31

Snack products	**Glycemic Index**
Tofu-based frozen dessert	115 ± 14
Pretzels	83 ± 9
Puffed rice cakes	82 ± 10
Jelly beans	80 ± 8
Licorice	78 ± 11
Pirate's Booty	70 ± 5
Angel food cake	67
Popcorn	65 ± 5
Water crackers	63 ± 9
Ice cream	62 ± 9
Potato chips	56 ± 3
Snickers bar	51
Milk chocolate, Dove	45 ± 8
Corn chips	42 ± 4

Low-fat yogurt	33 ± 3
M&Ms, peanut	33 ± 3
Dark chocolate, Dove	23 ± 3
Greek-style yogurt	12 ± 4
Hummus	6 ± 4

Choose high-fiber carbohydrates. High-fiber foods are one of your best longevity weapons. Years of research have found that the more fiber you eat, the better for your health and weight. How does dietary fiber fight fat? First, it helps regulate the appetite hormone ghrelin, which tells you that you are hungry. Ghrelin levels are often out of whack in people with a high BMI so you always feel hungry no matter how much you eat. New research shows that high ghrelin levels not only make you feel hungrier, they also increase the desire for high-calorie foods compared to low-calorie fare, so it's a double whammy. But fiber can help. A 2009 study showed that eating a diet high in fiber helped balance ghrelin levels, which can turn off the constant hunger and reduce the appeal of high-calorie foods that kill you early. Second, no matter how much you weigh, eating fiber helps you feel full longer so you don't get the munchies an hour after you eat. Third, fiber slows the absorption of food into the bloodstream, which helps balance your blood sugar and lowers the risk for diabetes. In fact, fiber takes so long to be digested by your body, a person eating 20–35 g fiber a day will burn an extra 150 calories a day or lose 16 extra pounds a year.

These three things alone can go a long way in helping you live longer. Fiber-friendly foods boast a number of other health benefits as well, including:

- Reducing cholesterol
- Keeping your digestive tract moving
- Reducing high blood pressure
- Reducing the risk of cancer

Experts recommend eating 25–35 g fiber a day, but research shows that most adults fall far short of that. So how can you boost your fiber intake? Eat more high-fiber brain healthy foods like fruits, vegetables, legumes, and whole grains. Here are the fiber contents of some brain healthy foods. Try to include some of the foods on this list at every meal or snack.

Food	Grams of Fiber
Kidney beans (1 cup canned)	16.4
Split peas (1 cup cooked)	16.4
Lentils (1 cup cooked)	15.6
Black beans (1 cup canned)	15.0
Garbanzo beans (1 cup canned)	10.6
Peas (1 cup frozen cooked)	8.8
Raspberries (1 cup)	8.0
Blackberries (1 cup)	7.6
Spinach (1 cup cooked)	7.0
(1 cup raw)	0.7
Brussels sprouts (1 cup cooked)	6.4
Broccoli (1 cup cooked)	5.6
Pear (1 medium w/skin)	5.1
Sweet potato (1 medium baked)	4.8
Carrots (1 cup cooked)	4.6
(1 medium raw)	2.0
Blueberries (1 cup)	3.5
Strawberries (1 cup)	3.3
Apple (1 medium w/skin)	3.3
Banana (1 medium)	3.1
Orange (1 medium)	3.1
Asparagus (1 cup cooked)	3.0
Grapefruit (½ medium)	2.0
Avocado (1 ounce)	1.9
Whole wheat bread (1 slice)	1.9
Walnuts (7 whole)	1.9
Plums (2 medium)	1.8

Peach (1 medium with skin)	1.5
Tomato (½ cup fresh)	1.5
Cherries (10 large)	1.4
Oatmeal (¾ cup cooked)	0.8
Almonds (6 whole)	0.8

Source: Adapted from U.S. Department of Agriculture, Agricultural Research Service, 2004. USDA Nutritional Nutrient Database for Standard Reference, Release 17.

Steer clear of bad carbohydrates. These are carbohydrates that have been robbed of any nutritional value, such as simple sugars and refined carbohydrates like muffins, scones, cakes, cookies, and other baked goods. If you want to live without cravings, eliminate these completely from your diet. I like the old saying "The whiter the bread, the faster you are dead."

Sugar is not your friend. We have often heard of sugar being called "empty calories." In fact, it is so damaging to your brain and body that I call it anti-nutrition or toxic calories. Sugar increases inflammation in your body, increases erratic brain cell firing, and sends your blood sugar levels on a roller-coaster ride. Moreover, new research shows that sugar is addictive and can even be *more* addictive than cocaine.

That helps explain why we eat so much of it. Americans consume an average of 22.2 teaspoons of sugar a day, which adds up to 355 calories a day. That's an increase of 19 percent since 1970.

Table sugar isn't the only culprit making you fat. Research shows that high-fructose corn syrup (HFCS), which is found in many sodas and accounts for as much as 40 percent of the caloric sweeteners used in the United States, is even more fattening than table sugar.

HFCS and sugar went head to head in a 2010 study from researchers at Princeton University. Compared with rats that drank water sweetened with table sugar, rats that drank water sweetened with HFCS gained significantly more body weight, including more fat around the belly, even though they consumed the same number of calories.

Every single one of the rats drinking the HFCS became obese. After six months, the rats guzzling the HFCS showed signs of a dangerous condition known in humans as the metabolic syndrome, including weight gain, abdominal fat, and high triglycerides.

Put down the sodas and HFCS, *now*!

A lot of people ask me, "Isn't it okay to have sweets in moderation?" Personally, I don't agree with the people who say "Everything in moderation." Cocaine or arsenic in moderation is not a good idea. Sugar in moderation triggers cravings. The less sugar in your life the better your life will be. Reach for a banana or an apple instead.

Cutting down on the sweet stuff is a good start, but sugar lurks in a lot of other processed foods as well like ketchup, barbecue sauce, and salad dressing. Start reading food labels. At first, you might feel like you're reading a foreign language—sorbitol, maltose, maltodextrin, galactose: these are just some of the many names for sugar used on food labels.

RULE 5. LIMIT FAT CONSUMPTION TO HEALTHY FATS, ESPECIALLY THOSE CONTAINING OMEGA-3s

Healthy fats are important to a good diet because the solid weight of the brain is 60 percent fat! The one hundred billion nerve cells in your brain need essential fatty acids to function. Focus your diet on healthy fats, especially those that contain omega-3 fatty acids, found in foods like salmon, tuna, mackerel, avocados, walnuts, and green leafy vegetables.

How omega-3 fatty acids help you get thinner, smarter, and happier. The two most studied omega-3 fatty acids are eicosapentaenoic acid (EPA) and docosahexaenoic acid (DHA). DHA makes up a large portion of the gray matter of the brain. The fat in your brain forms cell membranes and plays a vital role in how our cells function. Neurons are also rich in omega-3 fatty acids. EPA improves blood flow, which boosts overall brain function.

Low levels of omega-3 fatty acids have been associated with depression, anxiety, obesity, ADD, suicide, and an increased risk for Alzheimer's disease and dementia. There is also scientific evidence that low levels of omega-3 fatty acids play a role in substance abuse, and I would argue that *overeating is a form of substance abuse.*

Boosting omega-3 fatty acids in your diet is one of the best things you can do for your weight, mood, brainpower, and longevity. In a fascinating 2009 study in the *British Journal of Nutrition,* Australian researchers analyzed blood samples from 124 adults (21 healthy weight, 40 overweight, and 63 obese), calculated their BMI, and measured their waist and hip circumference. They found that obese individuals had significantly lower levels of EPA and DHA compared with healthy-weight people. Subjects with higher levels were more likely to have a healthy BMI and waist and hip measurements.

Research in the last few years has revealed that diets rich in omega-3 fatty acids may help promote a healthy emotional balance and positive mood in later years, possibly because DHA is a main component of the brain's synapses. A growing body of scientific evidence indicates that fish oil helps ease symptoms of depression. One twenty-year study involving 3,317 men and women found that people with the highest consumption of EPA and DHA were less likely to have symptoms of depression. People in Japan have the lowest levels of depression, and they eat the most fish. North Americans have high levels of depression, and we eat low amounts of fish.

There is a tremendous amount of scientific evidence pointing to a connection between the consumption of fish that is rich in omega-3 fatty acids and cognitive function. A Danish team of researchers compared the diets of 5,386 healthy older individuals and found that the more fish in a person's diet, the longer the person was able to maintain their memory and reduce the risk of dementia. Dr. J. A. Conquer and colleagues from the University of Guelph in Ontario, Canada, studied the blood fatty acid content in the early and later stages of dementia and noted low levels when compared with healthy people.

Eating fish also benefits cognitive performance. A study from Swedish researchers that surveyed nearly five thousand fifteen-year-old boys found that those who ate fish more than once a week scored higher on standard intelligence tests than teens who ate no fish. A follow-up study found that teens eating fish more than once a week also had better grades at school than students with lower fish consumption. Other benefits of omega-3 fatty acids include improving attention in people with ADD and reducing the risk for psychosis.

FOODS HIGH IN OMEGA-3 FATTY ACIDS

- Anchovies
- Broccoli
- Brussels sprouts
- Cabbage
- Cauliflower
- Cod
- Flaxseeds
- Halibut
- Mackerel
- Salmon, wild
- Sardines
- Scallops
- Shrimp
- Snapper
- Soybeans
- Spinach
- Tofu
- Trout
- Tuna
- Walnuts

Eliminate bad fats. While healthy fats enhance brainpower and help you lose weight, bad fats drain your brain. Eating too many saturated fats or trans fats, also known as "Frankenfats," contributes to obesity and cognitive decline. Trans fats are used to help foods have a longer shelf life and are found in margarine, cakes, crackers, cookies, and potato chips. They decrease *your* shelf life!

RULE 6. EAT NATURAL FOODS OF MANY DIFFERENT COLORS TO BOOST ANTIOXIDANTS

This means you should eat all the colors of the rainbow. Eat blue foods (blueberries), red foods (pomegranates, strawberries, raspberries, cherries, red bell peppers, and tomatoes), yellow foods (squash, yellow bell peppers, small portions of bananas, and peaches), orange foods (oranges, tangerines, and yams), green foods (spinach, broccoli, and peas), purple foods (plums), and so on.

This will boost the antioxidant levels in your body and help keep your brain young. Several studies have found that eating foods rich in antioxidants, which include many fruits and vegetables, significantly reduces the risk of developing cognitive impairment.

Blueberries are very high in antioxidants, which has earned them the nickname "brain berries" among neuroscientists. In laboratory studies, rats that ate blueberries showed a better ability to learn new motor skills and gained protection against strokes. That is not all. In one study, rats that ate a diet rich in blueberries lost abdominal fat, lowered cholesterol, and improved glucose levels. Similar studies showed that rats that consumed strawberries and spinach also gained significant protection.

Eating fruits and vegetables from the rainbow, along with fish, legumes, and nuts is part of what is known as the Mediterranean diet. Research has found that eating a Mediterranean diet can make you not only happier but smarter too. A series of studies from Spanish researchers revealed that adherence to this type of eating plan helps prevent depression. A team of scientists in Bordeaux, France, concluded that a Mediterranean diet slows cognitive decline and reduces the risk for dementia.

Of course, eating from the rainbow does *not* mean indulging in Skittles or jelly beans.

FRUITS AND VEGETABLES WITH HIGH ANTIOXIDANT LEVELS

- Acai berries
- Avocados
- Beets
- Blackberries
- Blueberries
- Broccoli
- Brussels sprouts
- Cherries
- Cranberries
- Kiwis
- Oranges
- Plums
- Pomegranates
- Raspberries
- Red bell peppers
- Red grapes
- Spinach
- Strawberries

RULE 7. COOK WITH BRAIN HEALTHY HERBS AND SPICES

Cutting calories while keeping taste and pleasure high is made much easier by the creative use of spices. Spices not only increase the flavor of foods without adding salt, but most of them have wonderful de-aging and health-boosting properties. It only takes a teaspoon of spices to add a powerful, concentrated antioxidant punch to your health profile. Try these ten brain healthy spices to help keep you young:

Turmeric Found in curry, turmeric contains a chemical that has been shown to decrease the plaques in the brain thought to be responsible for Alzheimer's disease.

Saffron In three studies, a saffron extract was found to be as effective as antidepressant medication in treating people with major depression.

Sage Sage has very good scientific evidence that it helps boost memory.

Cinnamon Cinnamon has been shown to help attention and it helps regulate blood sugar, which decreases cravings. Also, cinnamon is a natural aphrodisiac for men—not that most men need much help.

Basil This potent antioxidant improves blood flow to the heart and brain and has anti-inflammatory properties that offer protection from Alzheimer's disease.

Thyme Supplementing the diet with thyme has been shown to increase the amount of DHA—an essential fatty acid—in the brain.

Oregano Dried oregano has thirty times the brain-healing antioxidant power of raw blueberries, forty-six times more than apples, and fifty-six times as much as strawberries, making it one of the most powerful brain cell protectors on the planet.

Garlic Garlic promotes better blood flow to the brain and killed brain cancer cells in a 2007 study.

Ginger Can ginger make you smarter? A study that combined ginger with ginkgo biloba suggests that it does. Ginger root extract may also be helpful in the treatment of Parkinson's disease and migraine headaches.

Rosemary A 2006 study reported that rosemary diminishes cognitive decline in people with dementia.

MORE WAYS TO HELP YOU AND YOUR BRAIN LIVE LONGER

LIMIT CAFFEINE

Most of us associate caffeine with coffee, but it can also be found in tea, dark sodas, chocolate, energy drinks, and pep pills. If your caffeine intake is limited to one or two normal-size cups of coffee or two or three cups of tea a day, it probably is not a problem. But any more than that can cause problems.

Why?

Caffeine restricts blood flow to the brain. Anything that compromises blood flow leads to premature aging.

Caffeine dehydrates the brain and body. This makes it harder to think quickly. Remember, your brain is 80 percent water and needs adequate hydration.

Caffeine interferes with sleep. Sleep is essential for good brain health, appetite control, and skin rejuvenation. Caffeine disrupts sleep patterns because it blocks adenosine, a chemical that tells us when it is time to hit the hay. When this chemical is blocked, we tend to sleep less, which leads to sleep deprivation. And when we aren't getting enough sleep, we feel like we absolutely must have that cup of joe in the morning in order to jump-start our day.

Caffeine can be addictive in high amounts. When you try to kick the habit, you are likely to experience withdrawal symptoms, including severe headaches, fatigue, and irritability.

Caffeine can accelerate heart rate and raise blood pressure. In some people, drinking too much caffeine leads to a temporary spike in blood pressure and a racing heart.

Caffeine can give you the jitters. Ingesting more caffeine than you normally do can leave you feeling jittery and nervous.

Caffeine increases muscle tension. Tight muscles have been linked to caffeine intake.

Caffeine can cause an upset stomach. Gastrointestinal troubles are common with excessive caffeine use.

Caffeine can elevate inflammatory markers. Two studies showed that 200 mg caffeine (equivalent to two cups of coffee) raised homocysteine levels, a marker for inflammation and heart disease.

Caffeine can interfere with fertility. Pregnant women should be careful with caffeine because it has been associated with premature births, birth defects, inability to conceive, low birth weight, and miscarriage.

To be fair, there are also a number of studies suggesting that coffee can be helpful for you. It has been shown to decrease the plaques that cause Alzheimer's disease, lower the risk for Parkinson's disease, and lower the risk of colon cancer and diabetes. It may be other substances in the coffee, not just the caffeine, that are actually helpful, and decaffeinated varieties may give you the benefits without the troubles noted above. A Harvard University study found that those drinking decaffeinated coffee also showed a reduced diabetes risk, though it was half as much as those drinking caffeinated coffee. Another study, however, found that caffeine reduced insulin sensitivity and raised blood sugar—both bad news for you. Watch how your body responds to caffeine and keep it to a minimum.

EAT SUPER BRAIN FOODS

Unless you've been living in isolation for the past ten years, you've probably heard about antioxidants, which neutralize the production of free radicals in the body; free radicals, if left to themselves, can do any number of dirty tricks to your body. Cancer, aging too fast, and lowered immunity to all diseases are just a few things that can happen unless we load up on our superhero foods that are full of antioxidants. Foods rich in antioxidants include a variety of fruits and vegetables.

Scientists have recently reported the first evidence that eating blueberries, strawberries, and acai berries may help the aging brain stay healthy in a crucial but previously unrecognized way. Their study concluded that berries, and possibly walnuts, activate the brain's natural "housekeeper" mechanism, which cleans up and recycles toxic proteins linked to age-related memory loss and other mental decline. We've known from past studies that old lab rats fed for two months on diets containing 2 percent high-antioxidant strawberry, blueberry, or blackberry extract showed a reversal of age-related deficits in nerve function, memory, and learning behaviors. But in this new study, researchers found out that berries help the aging microglia (think brooms that have gone soft and don't work well anymore) perk up and clean up the bad brain debris that leads to aging.

An apple a day may indeed keep the doctor away. Scientists are reporting that consuming a healthful antioxidant substance in apples extends the average life span of test animals—by 10 percent. The researchers found that apple polyphenols not only prolonged the average life span of fruit flies but also helped preserve their ability to walk, climb, and move about. In addition, apple polyphenols reversed the biological markers for age-related deterioration and approaching death. In another study, women who often ate apples had a 13–22 percent decrease in the risk of heart disease.

I've mentioned studies on just two kinds of fruit here: berries and apples. There are hundreds of studies on the antioxidant, antiaging values of hundreds of fruits and vegetables. Mom was right to urge us to eat them.

When it comes to antioxidants, consider Tana's advice to Tamara: "Turn your fridge into a rainbow." Eating fruits and vegetables of many different colors will ensure that you are getting a wide variety of antioxidants to nourish and protect your brain. (Tana's cookbook *Eat Healthy with the Brain Doctor's Wife* has fabulous recipes that make use of the best in brain foods.)

REDUCE SALT AND INCREASE POTASSIUM

Reducing salt, especially the sort of salt found in processed foods, is another important health measure. Using sea salt, because it is rich in minerals and more flavorful than regular table salt, to season your own fresh food is a good idea. Canned vegetables, like corn, tend to have much more salt than you'd use if you cooked up fresh or frozen corn and lightly sprinkled it with sea salt. Most people end up using less of it than regular salt because it packs such a flavor punch. Simply buying more fresh or frozen vegetables instead of canned ones can help cut your sodium intake significantly. The same is true for preflavored or marinated meats. Avoid these and make your own. It won't take but a few seconds to whip up a marinade that tastes a hundred times better and has significantly less salt than prepackaged, preseasoned meats.

Also consider adding potassium to your diet. A recent study found that eating twice as much potassium as sodium can cut the risk of dying from heart disease in half. A 1997 study in the *Journal of the American Medical Association* that reviewed the results of thirty-three clinical trials found that individuals who took potassium supplements lowered their blood pressure. Foods high in potassium include bananas, spinach, honeydew melon, kiwi, lima beans, oranges, tomatoes, and all meats.

ALLOW SNACKING

If anybody has ever told you to avoid snacking throughout the day, don't listen! Going too long without eating can wreak havoc on your brain function and make your blood sugar levels drop too low. Low blood sugar levels are associated with poor impulse control and irritability. It can also cause emotional stress in some people.

Eating approximately every 2.5–3 hours throughout the day can help balance your blood sugar. This isn't a license to gorge all day long. When snacking, opt for low-calorie foods and include a balance of protein, complex carbs, and good fats, if possible. Personally, I love to snack. Since I travel frequently, I've learned to pack brain healthy snacks for the road. Otherwise, I am tempted to grab candy bars from the airport gift shop. One of my favorite low-calorie snacks is dried fruits, without any added sugar or preservatives, and fresh raw vegetables. I add a few nuts or some low-fat string cheese to balance out the carbohydrates in the fruits and vegetables with a little protein and fat. Be wary when buying dried fruits and vegetables—many brands add sugar, preservatives, or other ingredients, which renders them less than healthy. Read the food labels. Look for brands that don't add anything.

Here are a few more of my favorite midmorning or midafternoon snacks:

- Chopped veggies and hummus
- Fresh guacamole and red bell peppers
- Celery with almond butter
- Apple slices or a banana with almond butter
- Unsweetened yogurt and blueberries with a little stevia
- Deviled eggs with hummus (discard the yolk)
- Turkey and apple slices with a few almonds
- Steamed edamame
- Fresh guacamole on sprouted grain toast
- Homemade turkey jerky

RECOGNIZE FOOD ALLERGIES

Many people know that food allergies can cause hives, itching, eczema, nausea, diarrhea, and, in severe cases, shock or constriction of the airways, which can make it difficult to breathe and can be fatal. But can certain foods and food additives also cause emotional, behavioral, or learning problems? You bet. These types of reactions are called "hidden" food allergies, and they could be hampering your efforts for a better body. Tana and her sister, Tamara, are both classic examples of this. Both react severely to gluten and to dairy. I had one patient who found he got extremely angry after eating MSG (monosodium glutamate). It is always a good idea to ask yourself what you ate before experiencing negative emotions or off-kilter physical symptoms.

As far as food allergies go, the most common culprits are peanuts, milk, eggs, soy, fish, shellfish, tree nuts, and wheat. These eight foods account for 90 percent of all food-allergic reactions. Other foods commonly associated with allergies include corn, chocolate, tea, coffee, sugar, yeast, citrus fruits, pork, rye, beef, tomato, and barley.

Physical symptoms that might tip you off to a food allergy or sensitivity include dark circles under the eyes, puffy eyes, headaches or migraines, red ears, fatigue, joint pain, chronic sinus problems (congestion or runny nose), or gastrointestinal issues. Behavioral problems that can be caused by foods include aggression, sleep problems, lack of concentration, and changes in speech patterns (turning into a motormouth or slurring words).

When a food allergy or food sensitivity is suspected, a medical professional may recommend an elimination diet such as the "gorilla diet" that Tana recommended for Tamara. An elimination diet removes all common problem foods for a period of one or more weeks. These diets aren't easy to follow because they're very restrictive. However, if you've been miserable from potential food reactions for a long time, the fact that you may begin to feel better soon (as in Tamara's case) helps with the motivation needed to stay on the restrictive diet for a while. After the initial diet period, potential allergens are reintroduced one by one. Foods that cause abnormal behaviors or physical symptoms should be permanently eliminated from the diet. Working with a nutritionist may make a big difference. For a long while, if Tamara ate so much as a piece of bread, her miseries would return. But now that she's doing so much better and healing well, she can get away with a small amount of her husband's incredible Spanish rice now and then. But it is a treat—she eats a small portion and she doesn't eat it often.

HORMONES, ORGANICS, AND SAFE LISTS . . . OH MY!

There is rising concern about pesticides, used on plants for food, causing endocrine disruption, meaning that the residual pesticides appear to be changing hormone levels in our populations. This concern is compounded by the use of hormones in the dairy and beef industry in order to increase milk production and "beef up" the cattle. The most common problem is that foods now have too many "estrogenic" properties, causing some feminization in males, such as less sperm production, less sex drive, and prostate problems. In women, we are seeing a rise in early puberty and menopause. For these reasons, it is a good idea to choose hormone-free and organic foods whenever possible. Also, eating omega-3-rich foods, exercise, and consuming lots of fruits and vegetables will help flush excess estrogen out of your system.

The following lists from the Environmental Working Group contain the most contaminated "dirty dozen" foods (you should buy their organic counterparts instead) and those that are least contaminated. Keeping these lists handy while shopping can help stretch your budget, so you can use your dollars wisely.

TWELVE MOST CONTAMINATED FRUITS AND VEGETABLES

1. Celery
2. Peaches
3. Strawberries
4. Apples
5. Blueberries
6. Nectarines
7. Bell peppers
8. Spinach
9. Cherries
10. Collard greens and kale
11. Potatoes
12. Grapes

TWELVE LEAST CONTAMINATED FRUITS AND VEGETABLES

1. Onions
2. Avocado
3. Sweet corn (frozen)
4. Pineapples
5. Mango

6. Asparagus
7. Sweet peas (frozen)
8. Kiwi fruit
9. Bananas
10. Cabbage
11. Broccoli
12. Papaya

What about fish? On the one hand, nutritionists are encouraging us to eat fish, but they are also cautioning us about the mercury levels in some of the fish, which if eaten too often or in great quantities, could be cause for concern. For a downloadable pocket guide to the safest fish in your area of the country go to http://www.montereybayaquarium.org and click on "Seafood Watch." Here are a couple of general rules of thumb that are helpful: (1) the larger the fish, the more mercury it may contain, so go for the smaller varieties; and (2) from the safe fish choices, eat a fairly wide variety of fish, preferably those high in omega-3s. The following lists should be helpful.

"ABSOLUTELY AVOID" FISH LIST

(The following fish have been noted on various lists as having the highest amounts of mercury.)

Bluefin tuna
Bigeye tuna (also sold as ahi tuna)
King mackerel
Shark
Swordfish
Tilefish

SAFEST SEAFOOD CHOICES AND OMEGA-3s

(The following are the fish species listed as having the least amount of mercury, according to the National Resources Defense Council. They are ordered from highest to lowest in omega-3s per sensible serving.)

Fish	Omega-3s (grams)	Fat (grams)
Salmon, wild, Alaska, baked/broiled, 4 oz.	2.5	9.2
Whitefish, baked/broiled, 4 oz.	2.1	8.5
Mackerel (Atlantic), baked/broiled, 4 oz.	1.5	20
Trout (freshwater) baked/broiled, 4 oz.	1.3	7
Sardines, without skin, in water, 2 oz	1.3	7
Anchovies, canned in oil, drained, 2 oz.	1.2	5.5
Herring (Atlantic) baked/broiled, 2 oz.	1.2	6.5
Striped bass (farmed) baked/broiled, 4 oz.	1.1	3.4
Caviar (U.S. farmed), granular, 1 tbs.	1.05	2.9
Salmon, canned pink, drained 3 oz.	1	4.1
Squid, baked, 4 oz.	.8	5.3
Halibut (Alaska), baked/broiled, 4 oz.	.6	3.3
Sturgeon (U.S. farmed), baked/broiled, 4 oz.	.6	5.9
Sole (Pacific), baked/broiled, 4 oz.	.6	1.7
Crab, steamed, 4 oz.	.5	1.4
Catfish (U.S. farmed), baked/broiled, 4 oz.	.4	3.2
Scallops (bay farmed), steamed, 4 oz.	.4	1.6
Shrimp (U.S. farmed), steamed, 4 oz.	.4	1.2
Oysters (farmed), baked/broiled, 2 oz.	.3	1.2
Clams (farmed), steamed, 2 oz.	.2	1.1
Tilapia (U.S. farmed), baked/broiled, 4 oz.	.2	3
Abalone (U.S. farmed), cooked, 3 oz.	.06	.9

Source: ESHA Research Food Processor SQL, 2008.

BRAIN HEALTHY EATING ON A BUDGET

After the recession of recent years, many have discovered they have way too much month left at the end of their paycheck. How can one eat healthy and stay on budget too? Here are some tips from our online community participants:

- "I've become very freezer-happy lately as we've cut back on our food costs. If I see something about to go bad, I figure out a way to freeze it so we have minimal waste. During those times when

I'm stretching the groceries, I may run out of fresh produce, but I usually can hit up my freezer for a meal or two that has plenty of veggies in it, rather than living off canned food and pantry staples to get us through to the next paycheck."

- "When choosing organic, I spend $$ on the dirty dozen and buy non-organic from least contaminated list."
- "I buy frozen organic blueberries in bulk."
- "I try to buy in season from farmers' markets; then I'll freeze fresh produce if it is less expensive than frozen at that time."
- "I shop the Manager's Special section for great deals on fruits & veggies for juicing or using in smoothies or to put in the evening's dinner."
- "Though we don't eat much meat, I have friends who buy whole or half a cow or pig from a nearby farmer that doesn't use antibiotics or hormones. There's an initial investment for the deep freezer plus the bulk meat, but pays off in the end."
- "More and more people are raising their own chickens for eggs. I like to buy free-range eggs at the farmers' market or from a local farmer."
- "Take the time to learn how to store fruits & veggies properly so you get the optimal freshness out of them."
- "Freeze fruit for smoothies before it goes bad. Bananas often get ripe all at once, so I peel them, store them in plastic bag, and enjoy them either in smoothies or as treats."
- "Freezing nuts is also a good hint because they are cheaper in bulk but stay freshest in the freezer."
- "You can get the best prices on a lot of health products shopping online, especially if you buy in bulk. Chia seeds, nuts, and coconut oil are good for this. Health food stores will often discount bulk orders, too."
- "Make a Crock-Pot full of seasoned beans (pinto or black are wonderful) and season with spices, garlic and onions, then serve with salsa or pico de gallo over brown rice. You can top it with

jalapeños, Greek yogurt (tastes just like sour cream but with more protein, less calories), and avocado. You can mash leftover beans for use in healthy burritos on whole wheat tortillas for lunches. For a Southwest treat, serve a side of beans with huevos rancheros."

- "Join a local co-op."
- "Join a gardening co-op."
- "Ground turkey and larger bags of hormone-free chicken thighs and legs are cheap, healthy eats!"
- "Organic tilapia, bought in bulk, frozen is a fabulous fish that cooks in no time and is usually very affordable. Just sprinkle with a little of your favorite seasoning on one side, saute in a small amount of olive or coconut oil. Squeeze a slice of citrus on top and dinner is served. Great for fish tacos."

NUTRITIONAL GUIDES AND FOOD LABELS

When eating out always ask for the nutritional guide, which will help you make better decisions. Most restaurants now have these guides, and they will horrify you worse than *Piranha* 3D. Some appetizers like onion rings or potato skins can cost you more than 1,200 calories, more than half of your daily calorie allowance. When I was in a coffee house recently with a friend who was having trouble with her weight, I helped her discover that the latte she was about to order was 600 calories, a third of her daily calorie allowance for one drink. Shocked, she ordered green tea instead, which was virtually calorie free. I love saving money . . . and I love getting great value from my food.

Likewise, always read the food labels. I was on vacation recently with my son, Antony, and we were shopping together when he put a "healthy" protein bar in our cart. We then looked at the food label together and it had fourteen different types of sugar in it, along with artificial dyes and sweeteners! I told my son that the bar must have been mislabeled, because really it was an early-*death* bar. Read the labels. If you don't know what's in something, don't eat it. Would you ever buy something without knowing the cost? Be smart with your food and you will be smarter and younger for a longer period of time.

RIZ'S STORY

We started this chapter with the story of how simple changes in food made dramatic changes in my sister-in-law's life. This is also true for one of the physicians, Dr. Riz Malik, who works in our Reston, Virginia, clinic. Riz is an outstanding child psychiatrist who works very hard and has many grateful patients. One day he sent me an e-mail with a subject line that read "A Different Person." In the e-mail, Riz sent me the two photos you see below with a note saying, "Hey guys, I lost twenty-eight pounds in the last three months. The pictures show the difference. I just wanted to share."

I was so excited to see the change in Riz that I had to find out how he did it. He said it happened around the time my book *Change Your Brain, Change Your Body* was released. When a friend of his got diagnosed with high cholesterol and high blood pressure, it made him start to examine his own health and well-being.

After Riz came to the United States fifteen years ago at age twenty-five, he picked up the Western habits of eating lots of rich food loaded with saturated fat and drinking sugary beverages. He averaged four sodas (about 400 calories) and two fast-food meals (about 700 calories each) during the workday, bringing him to about 1,800 calories *before* he ate dinner. For lunch, he was addicted to eating a burger with mushrooms and Swiss cheese. He craved them every day. He also craved Indian foods with heavy curry sauces and lots of bread to go with them. (The bread was not whole wheat but rather made from white flour.) In addition, he also started to crave potato bread and was eating three to four slices a day with butter.

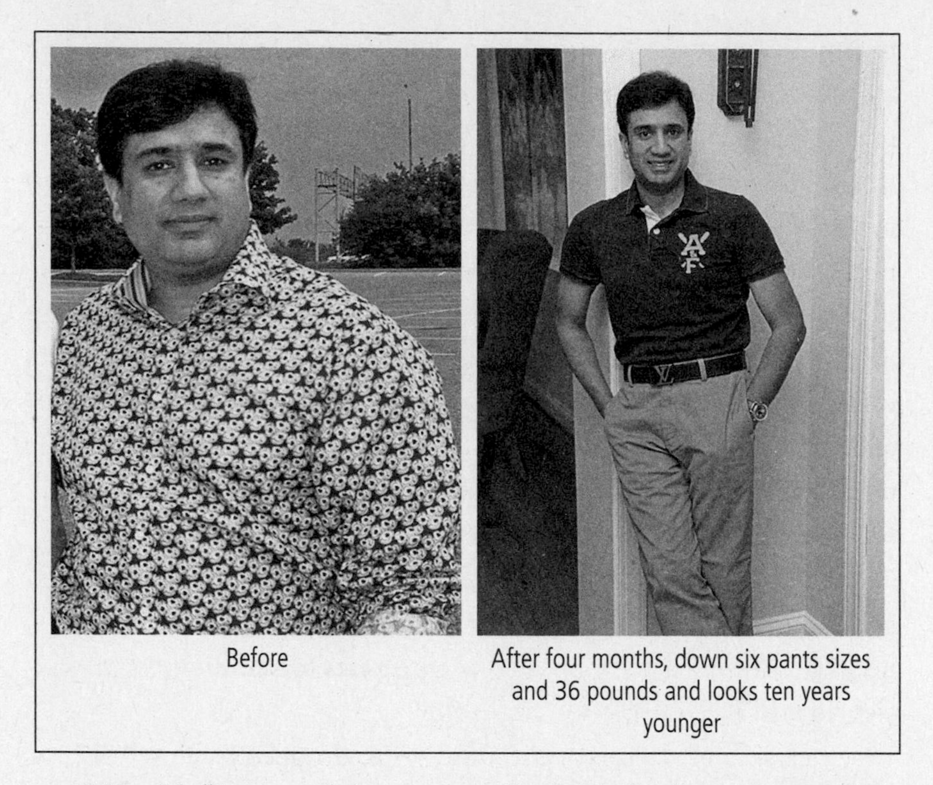

Before

After four months, down six pants sizes and 36 pounds and looks ten years younger

Riz said, "It seemed that eating all of this food was a way of rewarding myself for the hard work, the night calls, and the fact that I sometimes got overwhelmed with challenging psychiatric patients as I had to 'absorb' all the negativity, sadness, and mental turmoil my patients go through. So eating was a good way out. But now I feel that I was, in fact, penalizing myself by eating all those carb- and fat-loaded meals."

This is a key reminder, because people often believe they are "treating themselves" when they eat high-fat or -sugar foods, when in reality they are punishing themselves, treating their bodies poorly. Those high-fat meals weren't doing him any favors. In addition to pushing his weight to 183 and his BMI to 27, they gave him gastroesophageal reflux disease (GERD), also known as acid reflux. At forty years of age now, he thought that if he kept going in this direction, he would weigh about 250 pounds by the time he turned fifty-five.

Riz continued his story. "When I started looking at what I had done to myself, I thought, 'My God, this is not good.' I'm a physician and child psychiatrist, and here I am telling my patients and kids to eliminate sugar and eat healthy, but what am I doing?"

Riz realized he wasn't eating any vegetables or getting much fiber in his diet. Using the tips in our books, as well as other reputable nutrition sources, he started making changes to his eating habits. He ditched the sodas and switched to water with a little lemon. He started eating organic apples and blueberries, yogurt with fiber, skinless chicken and turkey breast, low-calorie whole grain bread, high-fiber lentils, and lots of veggies. He also started taking fish oil and a supplement called Attention Support, which supports attention and focus, and doing about ten to fifteen minutes of light exercise every day.

At first, the thought of giving up his burgers and comfort foods seemed almost impossible. "But after a while, you get used to the healthy food and start to like it." Now he looks forward to his blueberries, yogurt, and other brain healthy fare.

After four months, Riz lost 40 pounds, dropped his BMI to 22.7, and whittled his waistline from 36 to 30 inches. He weighs 148 pounds now, but weight loss isn't the only benefit from the changes he has made. "One of the most amazing things is my sleep has improved," he said. "I had GERD, and now that's gone. Plus, my level of alertness is better, and I'm more efficient at work."

Recently Riz sent this follow-up e-mail, which brought a big smile to my face, and I hope will inspire you to make the food choices today that will yield a younger, healthier "different person" in your mirror tomorrow.

Daniel,

I cannot thank you enough for guiding me in the right direction for a healthier lifestyle and a fit body with a "fit" mind.

For the last 4 months, I have maintained myself at 148 lbs, which is the ideal weight for me (down from 184 lbs in March 2009). I have learned how to manage my cravings, how to reward myself with "yummy foods" once in a while, and how to regulate my own metabolism. Of course you know that my waist went down from 36" to 30" at this point, and I love it.

My sleep is much better, my energy levels are good throughout the day and I feel more confident when talking to my patients about having optimal weight, a nice shape to my body, and most important, having the high self-esteem and confidence that results when you are in good shape and can take charge of your negative thoughts, unhealthy lifestyle, and negative temptations for unhealthy foods.

I knew that I had lost weight and I looked good and could enjoy better-fitting clothes and jeans, but I've had a question in my mind for the last 2 months: "My external appearance is good and I look fit and slim, what about the different parameters in terms of labs and other studies? Am I really healthy inside?"

So I went to my doctor last week, at Washington Hospital Center, and said, "Dr. R! Could you please run some blood tests on me? I have lost weight, felt good about it, but I want to make sure that I see the real proof that I have done things right rather than just being obsessed about my weight issues." So we did a bunch of tests and the results came back three days ago.

I will share them with you and won't mind sharing with others too.

In December of 2009, I got a regular physical and bloodwork and I compared that to the ones done last week.

Serum Cholesterol: Previous 209 (getting to a borderline risk) Now 156 (fantastic)

Serum LDL: Previous 109 (High) Now 82

Serum HDL: Previous 24 (Low) Now 49

Serum Triglycerides: Previous 302 Now 137 (Fantastic. A result of saying good-bye to Big Macs and Whoppers)

Random Serum Glucose: Previous 109 Now 70 Serum HBA1c

Even though I do not have diabetes, there is a family history in distant relatives, so I just wanted to get the test. It is 5.0 (Great)

Hemoglobin levels, vitamin D levels, complete blood count, electrolytes, was absolutely fine. And because I am over 40, I did a PSA [prostate-specific antigen test], and it was super normal and fine too.

The other area of my concern was my abnormally high diastolic blood pressure. It was always between 88 and 92 (which does qualify for borderline hypertension).

Ever since I can remember in the last 10–12 years, my blood pressure was always high in diastolic range. Having a strong family history of hypertension and heart attacks, I was always "nervous" about this. Guess what? In the last 3 encounters (two with my primary care physician and one with an ER nurse) in the last 8 weeks, my blood pressure has been consistently 115/75, 118/80, and 110/79, which is ideal.

Weight loss and reverting to a healthier lifestyle has been a blessing for me and it has helped me connect to my patients more and to offer them something beyond a pill to help them feel better.

I have been eating healthy, doing 15–20 minutes of exercise 4–5 days a week, getting to sleep on time, and taking fish oil and a multivitamin on a regular basis. It has helped a lot.

Please feel free to share my story with anyone it might encourage!

Regards,
Riz

As I was writing this book I was in our Reston clinic and, just looking at Riz, he clearly looked 10 years younger than before he lost the weight.

I WANT TO BE A TOUGH RED BELL PEPPER

I want to close this chapter with one of my favorite stories. I was recently on a walk with my wife and daughter near our home. At the time

Chloe was seven. She has red hair like her mother and usually speaks her mind. Even though it was a strenuous walk, Chloe did great. As we were nearing the end, her mother put her arm around her and said, "You are a tough cookie."

Immediately, Chloe looked up at her mother with an attitude and said, "I do not want to be a tough cookie. I want to be a tough red bell pepper."

Living in our house, Chloe knew about nutrition. One of her favorite snacks was sliced red bell peppers with fresh mashed avocados. She also knew that cookies and sugar made her feel bad, and having them increased the likelihood she would get herself into trouble. My hope for you is that you can be like Chloe and only want foods that serve you, not those that get you into trouble.

CHANGE YOUR AGE NOW: TWENTY BRAIN TIPS FOR MAKING FOOD YOUR FOUNTAIN OF YOUTH

1. Eat only food that serves you.

2. When eating out, ask the waiter to tailor your order to your health needs. Most are happy to help.

3. If you have multiple symptoms and digestive issues, you may have food allergies. Try the "gorilla diet" for a couple of weeks, eating only green foods, protein, and water, adding in new foods slowly to find the culprits causing your issues. The most common food allergy culprits are peanuts, milk, eggs, soy, fish, shellfish, tree nuts, and wheat, accounting for 90 percent of all food-allergic reactions.

4. Make your fridge into a rainbow. Stock it with all the most colorful fruits and vegetables in the deepest hues. Imagine gorgeous

dark spinach, red ripe tomatoes, crisp yellow bell peppers, sweet blueberries, and fresh cantaloupe lined up like a rainbow in your crisper. Getting your phytonutrients and antioxidants can be both beautiful and tasty!

5. Eating approximately every three to four hours throughout the day can help balance your blood sugar. Going too long without eating can wreak havoc on your brain function and make your blood sugar levels drop too low. Low blood sugar levels are associated with poor impulse control and irritability.

6. When eating out always ask for the nutritional guide, which will help you make better decisions. Also check labels, even on "health food" or "health bars," as packaging can be deceiving.

7. Eat with CROND in mind: calorie restricted, optimally nutritious, and delicious food. Begin making a list of foods that are packed with nutrition, that are delicious to you, and are low in calories. These foods will be your new best friends.

8. Junk food increases cravings. Be careful with the white powders: sugar, white flour, and salt.

9. Don't drink your calories. But *do* drink plenty of water, and white or green tea. (You can "spruce up" your water with a slice of citrus and a dash of stevia.)

10. Feed your brain good fats! Omega-3 fats found in foods like oily fish and flax or chia seeds are great for the brain. Olive oil is good for salads and light cooking; coconut, grape seed, and avocado oils are good for cooking at higher heat.

11. Rethink the idea of "rich food" as a reward for a hard day of work or accomplishing a task. If the rich or sugary food makes you fat and cranky or gives you heartburn, is it really such a great "reward"? Search for a win-win. Something nutritious and delicious that will leave you feeling better in an hour rather than worse.

12. Learn how to be creative with your food budget. The best fruits and veggies are the ones that are in season, and they are usually the ones on sale. Trim waste (and your waist) by freezing fruit and veggies before they go bad, and use them in smoothies or soups. Check out co-ops and other community resources for healthy food on a shoestring. Check the "dirty dozen" list of the most contaminated foods and budget money to buy these organically.

13. There is rising concern about hormones and toxins in foods that can change the healthy balance in your brain and body. When possible, buy hormone-free, antibiotic-free, free-range, grass-fed meats and dairy products.

14. Start your day with a good breakfast. One of the best ways to begin your day is with a fully loaded smoothie: Try a good-quality protein powder, some "green food" powder, and fresh veggies with a little fruit. You'll get hooked on the good mood and energy that comes from this powerful start to your day.

15. Watch for too much salt! Use it sparingly, and use sea salt at home. Avoid premarinated and canned foods that have much more sodium than is needed, raising blood pressure, which is bad for the brain, heart, and longevity.

16. Limit your caffeine intake. Some coffee may be beneficial for health, but many people are sensitive to its side effects. Moderation

is key. Green tea provides an energy boost without the caffeine-related jitters.

17. Spice it up. More and more studies show that spices are indeed the "spice of life"—in that many of them contain antioxidant and anti-inflammatory properties. Enjoy the good-for-you flavor of a variety of spices.

18. Eat three palm-sized servings of high-quality lean protein a day. Beans, fish, turkey, and chicken are especially good choices. Protein is a precursor to dopamine, epinephrine, and norepinephrine, which are critical for balancing mood and energy, metabolism and energy production, and improving cognitive performance.

19. Eating a diet that is filled with low-glycemic foods will stabilize your blood glucose levels and decrease cravings. Most low-GI foods are low in sugar and/or high in fiber or protein. The important concept to remember is that high blood sugar is bad for your brain and, ultimately, your longevity.

20. Supplements like fish oil, vitamin D, and specialized supplements that help control cravings or help with mood and focus may be helpful.

3

ANDY

GET STRONG TO LIVE LONG

According to one study, the number-one predictor of longevity is the amount of lean muscle mass on your body.

Tall, lean, smiling, and healthy at age sixty-four, exuding an unstoppable vibrancy that would make the Energizer Bunny envious, it is hard to believe that Dr. Andrew McGill began his life prematurely back in 1947. The expectations that he might actually live were so low that he wasn't named until he was four or five days old. But such is the strong will of one Andy McGill.

However, in 1999, when Andy was turning fifty-two, he had a wake-up call that would challenge his will to survive in a whole new way. It began when professor Andy McGill attended a seminar I gave at the University of Michigan about ADHD and our work with brain SPECT imaging. At the time, Andy's high school daughter, Katy, was struggling with ADHD. Curious about SPECT and wanting to make sure Katy had the best current diagnosis and follow-up treatment, Andy flew the whole family to California for a fun day of "having their heads examined" at one of our clinics. Andy heard me say in one of my presentations, "If your child has ADHD, look at the parents. More than likely he or she got it from one of you."

At that moment, Andy had looked at his wife, Kathe, suspiciously, thinking, "Well, it's obvious. Katy gets it from her mom." At the very same moment, Andy's wife was eyeing him, thinking, "Aha! My husband is the culprit!"

Originally, Andy and his wife's motivation for getting their brains scanned, alongside Katy, was to help their daughter feel less alone and more comfortable with the process. As it turned out, Katy, who had taken the summer off medication to prepare for the scan, had made such marked improvement that she was able to get off medication completely a few years later. However, Dr. McGill's scan told a different tale, and an alarming one.

BRAIN SPECT SCANS CAN TELL AN ALARMING STORY

When I first saw Andy's scan I wondered what I would tell him. It is not easy to tell a kind, educated man who had traveled more than halfway across the country to see you that his brain was damaged and looked much older than he was. His scan looked terrible. There was overall decreased activity in a Swiss cheese toxic-looking pattern. I had seen this pattern hundreds of times. It could be attributed to any number of causes: alcohol; drugs; environmental toxins, such as mold or organic solvents; infections; a lack of oxygen; or a significant medical problem like severe anemia or low thyroid.

When I asked Andy if he drank alcohol, he described what he felt was a pretty normal social drinking routine: a couple of drinks to wind down from a busy day of teaching business classes to college students, then a glass of wine with dinner, and finally, a nightcap before retiring for the evening. Since most "social drinkers" tend to fudge on the number of drinks they actually have, or the size of those drinks, I suspected that four drinks per day was probably the minimum daily drinking quota.

"How long have you been drinking?" I asked.

"Well, as a young man I was newspaper business editor for the *Miami Herald* and the *Detroit News,* so I was in a culture where smoking and

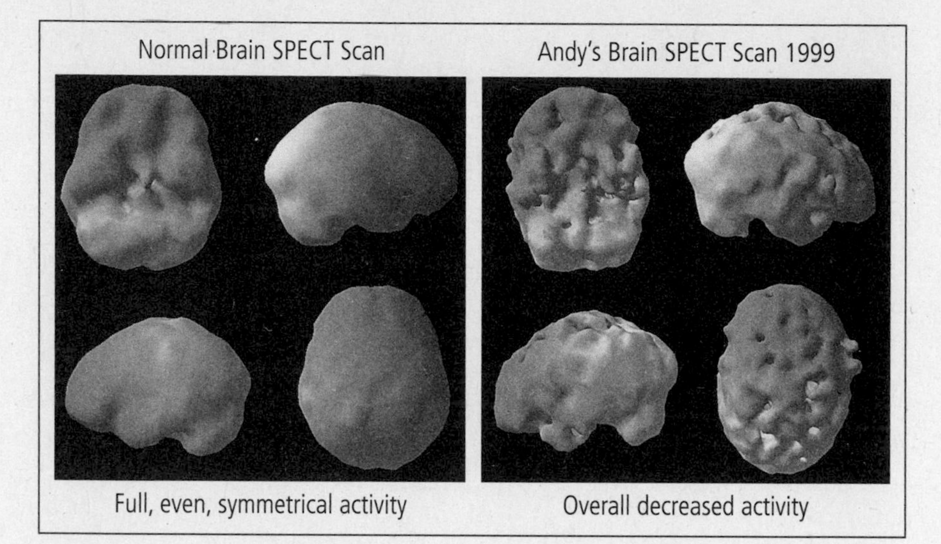

Full, even, symmetrical activity

Overall decreased activity

drinking went hand in hand with the job. I quit smoking back in the seventies. But I've put in about forty years of good drinking, all together."

I cut to the chase and told him, firmly and clearly, the realities of his future if he continued to drink, such as early dementia, disease, death, and basically losing his good mind. He was also quite overweight and needed to revamp his diet and start exercising. But the first order of business was that he had to stop poisoning his brain on a daily basis with alcohol. Also, much to his chagrin, I had to tell him the news that Katy's ADHD most likely came from her father's side of the brain tree.

Andy thanked me and flew back to Ann Arbor with his family. I had no idea whether or not he'd take my advice to heart (or should I say "to brain"?).

Then, about a year and a half later, our office received a call from Dr. McGill, who wanted to fly out to the clinic again, this time alone (he didn't even mention this trip to his wife), to recheck his brain. When I saw his scan, I shook my head sadly. His scan was even worse than the previous year.

This time, for whatever reason, Andy somberly assimilated the truth on a deep and profound level.

NEW YEAR, NEW LIFE

Andy will never forget New Year's Eve of the new millennium, ushering in 2001. He was drinking a glass of wine with some buddies at a party. Glasses in hand, they all began talking about how they wanted to stop drinking. Andy said, "I'm going to do it. When I finish this glass of wine, I'll never drink again. This is my last one." Amazingly, the McGill "will" kicked in. He never took another drink. (His friends all began drinking again within a few months.) Emotionally ready and intellectually armed after seeing his SPECT scan and hearing my concern, Andy made that solemn commitment, and he knew he would keep this promise to himself. He wanted a new brain for the new decade.

In addition to giving up alcohol, Andy also gave up caffeine, knowing from our talks that too much caffeine constricts blood flow to the brain. Once he had these two issues under control, he started feeling better and decided to tackle the extra 100 pounds.

When he first went to see his local doctor, his weight was at 279 pounds. He exercised sporadically. He got his weight down to 208 pounds in 2002, but it jumped up to 289.5 pounds in 2006. At this time, Andy was attending medical school—at age fifty-nine! He was taking classes and seeing patients as part of his clinical program. One day, he found himself lecturing a diabetic patient about the importance of nutrition, losing weight, and exercise. On the way home he thought to himself, "Andy, what a hypocrite you are! When are you going to stop playing around with your own health?"

Andy hit a wall on November 1, 2006, and not unlike what happened to him the day he stopped drinking, he made a deep, committed vow to himself never to miss a day of exercising. He's kept that promise for over five years now without missing a single day. Ever. Through sickness, through hard days ("through rain, through hail, through black of night") the ol' McGill will keeps on ticking. Once Andy fell on the ice and he was in pain, but having been through medical training, he knew he'd not broken or sprained anything. He was just bruised and sore. So to the treadmill he went.

"I actually find that when I'm not feeling well, which is rare, if I'll just get on the treadmill for a while, I will almost always feel better afterward."

DON'T BREAK THE STREAK!

"What I've learned about myself, based on my history, is that once I break a streak, it's easier to break it again and I might not stick with the program so seriously or get back with a program again," Andy explained. "What happens when you allow yourself to take breaks or start giving yourself excuses for days off is that you set up a cognitive precedent." A man of his word, even to himself, Andrew's secret of quitting a bad habit and starting (and sticking with) a new one is simple: Just do it. And don't stop doing it. *Don't break the streak.*

"The secret to exercise is to make it happen first thing in the morning, no excuses," Andy says. "The day will get away from you and you'll rationalize not exercising all day long if you don't do it as soon as you get up."

Andy has also incorporated one of the most positive outlooks on exercise I've ever observed. "I'll tell you what I love about exercising in the mornings," he says with a wide grin. "I know that no matter what else happens in my day, I can guarantee a part of my day will be great. When I exercise, I feel good. Even if there are troubles or frustrations as the day wears on, I've ensured I have had at least one fabulous hour by keeping exercise as a regular, top priority." In truth, because we know that exercise elevates mood and keeps the energy up throughout the day, Andy is also ensuring that his mood level stays as high as possible all day long, come what may.

A YOUNGER BRAIN, A SHARPER MIND, AND A HAPPIER LIFE

Now curious to see if all the changes he'd made over the years had affected his brain, Andy called the clinic for another appointment in late 2010. To tell you the truth, and as I would tell Andy later, I was dreading this, worried that his scan might be bad or have even gotten worse. I didn't know at this point about the scope of the lifestyle changes he'd been implementing, and of course, he was now almost a decade older since that first awful scan was taken.

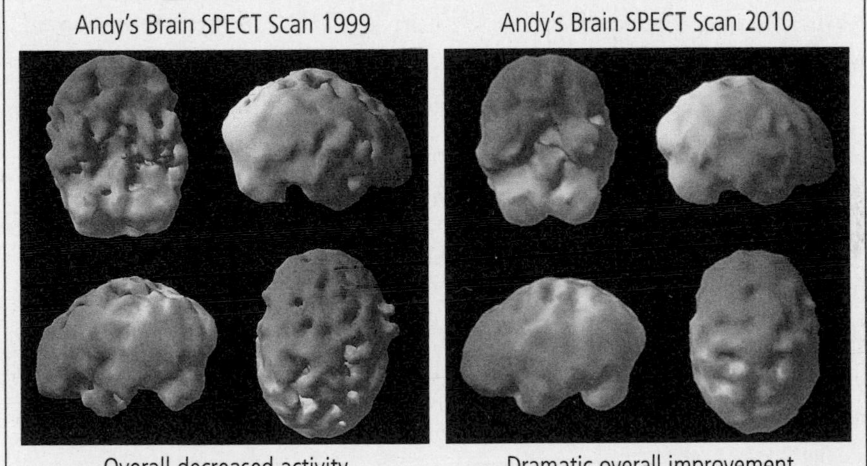

What I saw on Andy's new scan simply made my entire week. As we age, typically our brain looks older, less active. This guy's brain had aged backward! He was now the proud owner of a much younger, healthier brain than he had ten years before. I was elated to show Andy the scans and share the good news, the happy result of many years of omitting bad habits and steadily applying new ones.

When I asked Andy about his inspiration, he talked about his wife, Kathe, who had been doing her own swimming exercise routine—three times a week—for more than twenty years after being diagnosed with fibromyalgia, and had gotten back to a point of operating at 95 percent of her former mobility. She swims three days a week, in a short wetsuit after pouring hot water inside to warm her muscles, then doing a physical therapy routine of two dozen water exercises. Kathe has found that it is important to keep the muscles of fibromyalgia patients warmed up, and this way she enjoys the benefits of water exercise while keeping her muscles warm.

Today, Andy's energy level is great and stays steady all day. He doesn't feel run down or tired like so many his age. He also looks great and feels as intellectually engaged as when he was young, but now he has a wiser, more mature mind. He loves jazz and cooking, along with writing, some adjunct professorial duties, and some pro bono community service activities. He actually feels much more comfortable with students and people younger than himself than he does at the university retiree events. He feels decades younger than his biological age. His daughter, Katy, who once struggled in school, grew up to be a kindergarten teacher in San Diego and has a six-year-old, who happily runs his grandfather around Disneyland and Sea World when he visits. Andy McGill never takes these moments for granted, knowing full well that if he had not made vows to change his health habits ten years ago, he might not be here to see his grandson.

I recently received this encouraging and heartwarming e-mail from Andy, saying, "Along with the 2000 brain scan, it was your matter-of-fact warning, Dr. Amen, telling me that I would be in really sad shape by the time I reached sixty, unless I got serious about repairing the damage and got my brain healthy again. It scared me into action. That is when I became intellectually convinced. While it took me a few more months to become totally emotionally ready and quit drinking, you really saved my life and are responsible for me being as healthy as I am today. Never underestimate that. I may do the exercise. But you were the stimulant."

Needless to say, letters like this and lives changed for the better, like Andy McGill's, are why I love what I do.

ANDY'S IMPRESSIVE HEALTH STATS

Not only did Andy get a new, younger brain, he also got a new, younger body. When Andy began exercising, he was so out of shape and weighed so much that even walking at a very slow pace was tough. Today Andy is at an extreme level of fitness. For his morning routine now he runs on the treadmill for an hour, beginning at 3 mph at a 6 percent grade for the first 7.5 minutes, increasing gradually to 4.5 mph and 9 percent level. This routine gets his heart rate up to 125 beats per minute (bpm) from a resting rate of about 50 bpm. On Mondays, Wednesdays, and Fridays he does forty-five minutes of treadmill and then uses free weights for ten to fifteen minutes.

Andy reports, "I have now achieved a very, very high cardiac fitness level: the 99th percentile for men in my age group, comparatively—also at the 99th percentile for the age-fifty men's group, the 99th percentile for the age forty men's group. My regular doctor had to go down to the age thirty men's group to find a comparative level for me: There I am only in the 75th percentile!"

In addition, on Andy's last major fitness test, his VO_2 (volume of oxygen taken in during exercise) was 63 ml/kg/min (milligrams per kilogram per minute). "Normally, your VO_2 is expected to go down when you age," wrote Andy. "Mine went from 49, when I began my exercise streak in 2006, to 63. I'm told that is considered an extraordinary fitness achievement for my age. Achievements like this also help keep me motivated."

There are four major lessons in Andy's story.

1. Being able to see your own brain often creates brain envy and motivates people to start taking better care of it.

2. Alcohol is not your friend, especially lots of alcohol. The sooner you lose it or significantly decrease your intake, the better your brain and body will be.
3. Regular exercise can make a dramatic difference in how you look and feel.
4. Dedicate yourself to brain healthy habits and keep them going for a long, healthy life.

Andy

Before

After eleven years, down 100 pounds, with a better brain and cardiovascular system!

I used to exercise for my butt. Now I exercise for my brain.

—David Smith, M.D., founder of the Haight-Ashbury Free Clinics and my coauthor of *Unchain Your Brain*

This is one of my favorite quotes. David is a pioneer in the addiction treatment field, and together we wrote a terrific book, *Unchain Your Brain: 10 Steps to Breaking the Addictions That Steal Your Life,* on using the latest brain science to help people with addiction issues. As David learned more and more about creating a brain healthy life, his daily exercise routine became more about his brain than his body. Physical exercise is another fountain of youth for the health of your brain. Regular exercise will help you look younger, be trimmer, feel smarter, and boost your mood all at the same time.

Certainly, it is no secret that our society has shifted to a sedentary lifestyle where most of us spend our days sitting—working on computers, watching TV, and driving. The problem is that a lack of physical activity robs the brain of optimal function and is linked to obesity, higher rates of depression, a greater risk for cognitive impairment . . . and worse.

Physical inactivity is the fourth most common preventable cause of death, behind smoking, hypertension, and obesity.

EXERCISE AND LONGEVITY

New research seems to be popping up every day proving that exercise will not only increase how long you live but also the quality of life in those years. Here's a sample of some of the findings.

1. THOSE WITH PEP IN THEIR STEP STAY YOUNGER

A recent study shows that after age sixty-five, one strong predictor of longevity is walking speed. Those who can still hoof it after age seventy-five have an even better chance of living even longer. An eighty-year-old man who clocks 1 mph has a 10 percent probability of reaching age ninety, while a woman of the same age walking at that pace has a 23 percent chance. Now let's assume this pair is hoofing it a little faster at a speed of 3.5 mph. Now the eighty-year-old man has an 84 percent probability of reaching age ninety, while a woman would have an 86 percent chance.

2. DO YOUR AEROBICS AND KEEP MORE BRAIN TISSUE!

If you are like me, you'd like to keep every iota of brain tissue you possibly can as you get older. Researchers found that exercise, particularly aerobic exercise, reduces brain tissue loss in aging adults.

3. BALANCE EXERCISES HELP YOU AGE GRACEFULLY

Gentle exercise, like yoga or tai chi, increases balance, which decreases falls, which decreases injury and complications leading to death.

4. ACTIVE SENIORS LOOK YEARS YOUNGER THAN THEIR COUCH POTATO FRIENDS

Exercising thirty minutes a day, five times a week, can make you look many years younger than your biological age. Researchers from the University of St. Andrews in Scotland found that aging in the form of loose skin on the neck and jowls are the most pronounced effect of not exercising. The forehead and eye area also tend to fatten more in inactive people.

5. RESISTANCE EXERCISE KEEPS YOU STRONGER TO LIVE LONGER

University of Michigan researchers published a study showing that after an average of eighteen to twenty weeks of progressive resistance training, an adult can add 2.42 pounds of lean muscle to their body mass and increase their overall strength 25–30 percent. This is significant because without proactive training, aging adults tend to lose muscle mass and strength. The study recommended that people over age fifty begin by using their own body weight to do squats, modified push-ups, lying hip bridges, or standing up out of a chair. (Also, tai chi, Pilates, and yoga employ many resistance exercises using your own body weight as well.) Then you can add weights in a progressive training program designed specifically for your age and fitness, preferably starting with a personal trainer who can teach you good exercise form with weights, how many reps to do, and when it is time to up your weights.

6. THE GREATER YOUR MUSCLE STRENGTH, THE LESS RISK OF ALZHEIMER'S

According to research done at Rush University Medical Center in Chicago, individuals with weaker muscles appear to have a higher risk for Alzheimer's disease and declines in cognitive function over time. Those at the 90th percentile of muscle strength had about a 61 percent reduced risk of developing Alzheimer's disease compared with those in the 10th percentile. Overall, the data showed that greater muscle strength is associated with a decreased risk of developing Alzheimer's disease and mild cognitive impairment. It also suggests that a common but yet-to-be-determined factor may underlie loss of muscle strength and cognition in aging.

7. THOSE WHO EXERCISE SLOW DOWN THEIR BIOLOGICAL CLOCK

Exercise improves telomere maintenance by increasing the activity of the enzyme telomerase that builds and repairs telomeres. Telomeres are the part of your chromosomes that control aging. They represent your biological clock. When you are young your telomeres are longer, and they progressively shorten with age. But the rate at which that shortening occurs is directly influenced by lifestyle choices. So at any age, healthier individuals have longer telomeres than their unhealthy counterparts.

There are so many other benefits to regular exercise. Here are a few more of them.

Handle stress better. Working out helps you manage stress by immediately lowering stress hormones, and it makes you more resistant to stress over time. Raising your heart rate through exercise also makes you a better stress handler because it raises beta-endorphins, the brain's own natural morphine. Increasing your ability to manage stress can keep you from polishing off a whole bag of chips when you are under a lot of pressure.

Eat healthier foods. A 2008 study found that being physically active makes you more inclined to choose foods that are good for you, seek out more social support, and manage stress more effectively. Obviously, choosing brain healthy foods over junk food provides the foundation for lasting health. Creating a solid support network to encourage your new brain healthy habits can help you stay on track.

Get more restful sleep. Engaging in exercise on a routine basis normalizes melatonin production in the brain and improves sleeping habits. Getting better sleep improves brain function, helps you make better decisions about the foods you eat, and enhances your mood. Chronic lack of sleep nearly doubles your risk for obesity and is linked to depression and a sluggish brain.

Increase circulation. Physical activity improves your heart's ability to pump blood throughout your body, which increases blood flow to your brain. Better blood flow equals better overall brain function.

Grow more new brain cells. Exercise increases brain-derived neurotrophic factor (BDNF). BDNF is like an antiaging wonder drug that is involved with the growth of new brain cells. Think of BDNF as a sort of Miracle-Gro for your brain.

BDNF promotes learning and memory and makes your brain stronger. Specifically, exercise generates new brain cells in the temporal lobes (involved in memory) and the prefrontal cortex, or PFC (involved in planning and judgment). Having a strong PFC and temporal lobes is critical for successful weight loss.

A better memory helps you remember to do the important things that will help you stay healthy—for example, making an appointment with your physician to check your important health numbers, shopping for the foods that are the best for your brain, and taking the daily supplements that will benefit your brain type. Planning and judgment are vital because you need to plan meals and snacks in advance, and you need to make the best decisions throughout the day to stay on track.

The increased production of BDNF you get from exercise is only temporary. The new brain cells survive for about four weeks, then die off, unless they are stimulated with mental exercise or social interaction. This means you have to exercise on a regular basis in order to benefit from a continual supply of new brain cells. It also explains why people who work out at the gym and then go to the library are smarter than people who only work out at the gym.

Enhance brainpower. No matter how old you are, exercise increases your memory, your ability to think clearly, and your ability to plan. Decades of research have found that physical activity leads to better grades and higher test scores among students at all levels. It also boosts memory in young adults and improves frontal lobe function in older adults.

Getting your body moving also protects the short-term memory structures in the temporal lobes (hippocampus) from high-stress conditions. Stress causes the adrenal glands to produce excessive amounts of the hormone cortisol, which has been found to kill cells in the hippocampus and impair memory. In fact, people with Alzheimer's disease have higher cortisol levels than normal aging people.

Ward off memory loss and dementia. Exercise helps prevent, delay, and reduce the cognitive impairment that comes with aging, dementia, and Alzheimer's disease. In 2010 alone, more than a dozen studies reported that physical exercise results in a reduction in cognitive dysfunction in older people. One of them came from a group of Canadian researchers who looked at physical activity over the course of the lifetime of 9,344 women. Specifically, they looked at the women's activity levels as teenagers, at age thirty, at age fifty, and in late life. Physical activity as a teenager was associated with the lowest incidence of cognitive impairment later in life, but physical activity at *any* age correlated to reduced risk. This study tells me that it is never too late to start an exercise program.

Protect against brain injuries. Exercise strengthens the brain and enhances its ability to fight back against the damaging effects of brain injuries. This is so critical because brain injuries—even mild ones—can take the PFC offline, which reduces self-control, weakens your ability to say no to cravings, and increases the need for immediate gratification, as in "I must have that bacon cheeseburger *right this minute*!"

You don't have to lose consciousness to suffer from brain trauma. Even mild head injuries that do not typically show up on the structural brain imaging tests, such as MRIs or CT scans, can seriously impact your life and increase your risk for unhealthy behaviors. That is because trauma can affect not only the brain's hardware, or physical health, but also its software, or how it functions. Head injuries can disrupt and alter neurochemical functioning, resulting in emotional and behavioral problems, including an increased risk for eating problems and substance abuse.

Each year, two million new brain injuries are reported, and millions more go unreported. Brain trauma is especially common among people with addictions of all kinds, including food addiction. At Sierra Tucson, a world-renowned treatment center for addictions and behavioral disorders, our brain imaging technology has been used since 2009. One of the most surprising things the brain scans have shown, according to Robert Johnson, M.D., the facility's medical director, is a much higher than expected incidence of traumatic brain injury among their patients.

Get moving to get happier. Have you ever heard the term *runner's high*? Is it really possible to feel that good, just from exercise? You bet it is. Exercise can activate the same pathways in the brain as morphine and increases the release of endorphins, natural feel-good neurotransmitters. That makes exercise the closest thing to a happiness pill you will ever find.

Boost your mood. Physical exercise stimulates neurotransmitter activity, specifically norepinephrine, dopamine, and serotonin, which elevates mood.

Fight depression. In some people, exercise can be as effective as prescription medicine in treating depression. One of the reasons why exercise can be so useful is because BDNF not only grows new brain cells, but it is also instrumental in putting the brakes on depression.

The antidepressant benefits of exercise have been documented in medical literature. One study compared the benefits of exercise with those of the prescription antidepressant drug Zoloft. After twelve weeks, exercise proved equally effective as Zoloft in curbing depression. After ten months, exercise surpassed the effects of the drug. Minimizing symptoms of depression isn't the only way physical exercise outshined Zoloft.

Like all prescription medications for depression, Zoloft is associated with negative side effects, such as sexual dysfunction and lack of libido. Furthermore, taking Zoloft may ruin your ability to qualify for health insurance. Finally, popping a prescription pill doesn't help you learn any new skills. On the contrary, exercise improves your fitness, your shape, and your health, which also boosts self-esteem. It doesn't affect your insurability, and it allows you to gain new skills. If anyone in your family has feelings of depression, exercise can help.

I teach a course for people who suffer from depression, and one of the main things we cover is the importance of exercise in warding off this condition. I encourage all of these patients to start exercising and especially to engage in aerobic activity that gets the heart pumping. The results are truly amazing. Over time, many of these patients who have been taking antidepressant medication for years feel so much better that they are able to wean off the medicine.

Ease anxiety. Although the research on the effects of exercise on anxiety isn't quite as voluminous as the evidence on exercise and depression, it shows that physical activity of just about any kind and at any intensity level can soothe anxiety. In particular, high-intensity activity has been shown to reduce the incidence of panic attacks.

Boost your sexuality. Exercise helps to boost testosterone levels and makes you feel sexier. In addition, you look better, which also makes you feel and act more attractive. Even a few pounds or inches lost can make a big difference in how sexy we feel.

CHECK WITH YOUR DOCTOR BEFORE BEGINNING ANY EXERCISE PROGRAM.

BEST EXERCISES FOR YOUR BRAIN

Aerobic exercise, coordination activities, and resistance training have all been found to benefit the brain.

Get the most out of your aerobic exercise with burst training. If you want a higher-calorie burn, a faster-fat burn, a greater mood enhancer, and a better brain booster, try burst training. Also known as interval training, burst training involves sixty-second bursts at go-for-broke intensity followed by a few minutes of lower-intensity exertion. This is the type of workout I do, and it works. Scientific evidence says so. A 2006 study from researchers at the University of Guelph in Canada found that doing high-intensity burst training burns fat faster than continuous moderately intensive activities.

If you want to burn calories with bursts, do intense exercise, such as fast walking (walking as if you were late for an appointment), for thirty minutes at least four to five times a week. In addition, in each of these sessions, you are to do four one-minute bursts of intense exercise. These short bursts are essential to get the most out of your training. Short-burst training helps raise endorphins, lift your mood, and make you feel more energized. It also burns more calories and fat than continuous moderate exercise. Here is a sample of a heart-pumping thirty-minute workout with bursts:

SAMPLE BURST TRAINING WORKOUT

3 minutes	Warm-up
4 minutes	Fast walking (walk like you are late)
1 minute	Burst (run or walk as fast as you can)
4 minutes	Fast walking
1 minute	Burst
4 minutes	Fast walking
1 minute	Burst
4 minutes	Fast walking
1 minute	Burst
4 minutes	Fast walking
3 minutes	Cool down

If you can't devote an entire thirty minutes to an aerobic burst routine, don't throw in the towel. Research from Massachusetts General Hospital in Boston shows that just ten minutes of vigorous exercise can spark metabolic changes that promote fat burning, calorie burning, and better blood sugar control for at least an hour. For the 2010 trial, researchers looked at exercise-induced metabolic changes in people of varying fitness levels: people who became short of breath during exercise, healthy middle-aged individuals, and marathon runners.

All three groups benefited from ten minutes on a treadmill, but the fittest individuals got the biggest metabolic boost. This indicates that as you increase your fitness, your body will become more effective at burning fat and calories with exercise.

Boost your brain with coordination activities. Doing coordination activities—like dancing, tennis, or table tennis (the world's best brain sport)—that incorporate aerobic activity and coordination moves are the best brain boosters for all types of overeaters. The aerobic activity spawns new brain cells while the coordination moves strengthen the connections between those new cells so your brain can recruit them for other purposes, such as thinking, learning, and remembering.

What I really like about aerobic coordination activities is that many of them also work as burst-training sessions. For example, in tennis and table tennis, you give it your all during the point, then you have a brief rest period before the next point begins. It is the same with dancing, where you dance to the song and then take a short break.

In general, I recommend that all of us do some form of aerobic coordination activity at least four to five times a week for at least thirty minutes.

Have you typically avoided coordination activities because you have two left feet? This could be part of the reason why you have a hard time controlling yourself around food. That is because the cerebellum, which is the coordination center of the brain, is linked to the PFC, where judgment and decision-making occur. If you aren't very coordinated, it may indicate that you are not very good at making good decisions either. Increasing coordination exercises can activate the cerebellum, thereby improving your judgment so you can make better decisions.

Strengthen your brain with strength training. I also recommend adding resistance training to your workouts. Canadian researchers have found that resistance training plays a role in preventing cognitive decline. Plus, it builds muscle, which can rev your metabolism to help you burn more calories throughout the day. Extensive research shows that adding resistance training to a controlled-calorie nutrition program results in greater loss of body fat and more inches lost than diet alone.

Calm and focus your mind with mindful activities. Yoga, tai chi, and other mindful exercises have been found to reduce anxiety and depression and to increase focus. Although they don't offer the same BDNF-generating benefits as aerobic activity, these types of exercise can still boost your brain so you can improve your self-control and reduce emotional or anxious overeating.

"COME ON AND DANCE WITH ME!"

Eddie Deems has taught ballroom dance for seventy of his ninety-two years, which makes him one of the oldest—and most accomplished—hoofers in the Dallas–Fort Worth Metroplex. Lithe and graceful in his nineties, Eddie dresses in a handsome dark suit complete with ascot, looking every bit the professional dance instructor that he still is. Eddie is a living testimony to one of the world's best exercises for longevity: dancing!

Imagine going to the doctor, complaining of depression, and instead of giving you a prescription for Zoloft or Prozac, he hands you an Rx slip that says, "Take ten tango lessons and call me in two months." As far-fetched as that might sound, it could very well be a better answer than medication for many people with low-mood issues.

"We've become a nation of armchair dancers, mesmerized by *Dancing with the Stars* and *So You Think You Can Dance*," says Lane Anderson, author of an article in *Psychology Today*. "But research shows that getting your own groove on is more beneficial in improving social skills, lifting your spirits, even reversing depression.

"In a recent study at the University of Derby," Anderson wrote, "depressed patients given salsa-dancing lessons improved their moods significantly by the end of the nine-week, hip-swiveling therapy." Researchers found that the combination of the endorphin boost from exercise, along with the social interaction and forced concentration, lifted moods. I think it is safe to assume that the emotional boost of music, which calms and energizes the brain, also helped, along with the pride learning a new skill.

In a study from Germany, twenty-two tango dancers had lower levels of stress hormones and higher levels of testosterone. They also reported feeling sexier and more relaxed. Another study from the

University of New England showed that after six weeks of tango lessons, the participants showed significantly lower levels of depression than a control group who took no classes, and had similar results to a third group who took meditation lessons. Dance requires extreme focus or "mindfulness" and when the brain is deeply engaged at this level, negative thought patterns that lead to anxiety and depression are interrupted.

Using the body in physical, rhythmic movement also plays a part in opening people up on several levels. "Depressed patients tend to have a curved back, which brings the head down so it's facing the ground," said Donna Newman-Bluestein, a dance therapist with the American Dance Therapy Association. "Dancing lifts the body to an open, optimistic posture."

So grab your partner and do a little waltz around the kitchen, or go ahead and turn on "Dancing Queen" full volume (nobody's looking, right?) and get your groove on. What have you got to lose but a bad mood and a few pounds?

DR. JOE DISPENZA: HOW TO ENGAGE YOUR BRAIN IN A ONCE-AND-FOR-ALL DECISION

In addition to keeping a streak, another aspect that is unique to Andy's story was his "all in" approach to two life-altering decisions, from which there was no turning back. Not even a slip-up. It reminds me of that famous scene from *The Empire Strikes Back* in which Yoda tells Luke, "Do or do not. There is no try."

What enables people to make these strong vows to themselves that they simply don't break? Where there is no more "try," there is only "do."

Dr. Joe Dispenza, author of the book *Evolve the Brain: The Science of Changing Your Mind,* wrote: "I hope you've had this experience in your life where your intention, your focus and your will have all come into alignment." Dr. Dispenza has become a friend and refers many people to our clinics. I think he is on to something here. Ponder for a moment about vows you've made to yourself in the past, and think about what occurred in the times that you kept a vow and never broke from it. For some of you, perhaps, there was a moment where you said to yourself, "I will not be abused anymore" and broke free from an abuser, never to return. Or maybe it was the moment you chose your career path, knowing all the years of schooling and training ahead of you, knowing too that this was the course for you, so you signed up, went all in, and finished your degree. These were moments when your "intention, focus and will" collided to facilitate major change. I believe that intention, focus, and will stem from your PFC (the brain's supervisor) and your limbic system (emotional brain), working in tandem. Together they help you make the kind of deep, intrinsic, lasting vow to yourself to make radically better choices for your new, improved brain.

In a recent conversation with Dr. Joe Dispenza he shared more about the why and how of getting to an "all in" decision for your well-being, which could positively change the course of your life.

1. OBSERVE YOUR THOUGHTS

To begin the change process, Dr. Dispenza teaches others to become "metacognitive." This means that we step back from our thought patterns and observe them. We "think about our thinking patterns."

"'What if?' questions, when we start to speculate about new ways of thinking and being, open up a world of possibilities," he says. The PFC loves these kinds of questions, finds them stimulating. "Why not wake up every morning and begin the day by reminding yourself how you want to be and feel?" he asks. "And also remind yourself of who you don't want to be each day."

Dr. Dispenza points out that how you think creates how you feel (emotions). How you feel (your emotional state) creates a mood that, if left unchecked, eventually creates a temperament and, ultimately, your personality. Your personality, no doubt, ultimately affects your sense of reality. So much starts with a single thought. As the ancient proverb goes, "As a man thinks in his heart, so is he." But how do we really make up our minds to change? Dr. Dispenza says it has to do with creating a *firm intention* (with our PFC) that is strong enough to break old habits.

2. VISUALIZE YOUR NEW WANTED HABITS IN DETAIL

One way of creating new habits is to mentally practice. In one experiment, Dr. Dispenza points out that non-piano-playing people were taught to practice a series of finger movements on the piano for two hours a day, for five days. Another group of non–piano players was asked to mentally "practice" playing the piano (without moving their fingers) for the same amount of time. Brain scans showed that both groups showed the same pattern of new learning had taken place in the brain. Mental rehearsal, alone, changed the brain in the same way that actual practicing with the body did. Spending some time visualizing exactly how you'll spend your day in order to be healthier and live longer is a valuable mental tool in prepping you for that "all in" decision to change once and for all.

3. FEED YOUR BRAIN NEW EXPERIENCES TO CREATE NEW NEURAL NETS

Dr. Dispenza's work emphasizes the brain science truth that "neurons that fire together, wire together." The more that you feed your brain new experiences and new learning, the more those neurons fire; and the more you *repeat* similar experiences and layer similar knowledge, the more your neurons fire *and* wire together. Under heavy-duty microscopes, the neurons look very much like threads coming together to create a fishing net. In fact, this is what is called a neural net. Another way, then, to make a "once-and-for-all decision" is to feed your brain new experiences and new learning until your neurons "fire and wire together" to become new neuron nets, or new automatic thoughts and actions. For example, the more you read and study books like this one on brain health, the more you create connections of information that link together and begin to change your brain. The more you risk new experiences over time—like eating more fruits and vegetables every day or walking for thirty minutes a day—the more this becomes a good habit.

4. STUDY ROLE MODELS

Another way to help your brain change is to read and study those whose lives you'd like to emulate. Dr. Dispenza enjoyed Nelson Mandela's biography because it taught him how a noble man can suffer unjustly, for years, in a prison and then forgive and go on to great things with his life. The biography of the Wright Brothers is of great encouragement for someone risking a dream that seems unrealistic to many. Abraham Lincoln is a role model of honor, integrity, faith, and humor in a time of great crisis. I am sharing stories of real people in this book who have changed their brain, changed their life and their biological age, to help inspire you to believe that you too can change.

5. FIRM INTENTION

"The problem with the way most people make decisions to change is that they'll tell themselves, while lying on the couch with the remote, eating junk food and drinking beer, 'I'm going to get my act together and change tomorrow,'" explains Dr. Dispenza. "But the body is saying to the mind, 'Oh, relax. He doesn't really mean it. He always says this over and over but he never changes. Go ahead and have another potato chip and a swig of beer.'

But when you really truly make up your mind and are in a state of firm intention—that is, you absolutely *know* that you are going to follow through on a new way of thinking or acting, you can almost feel the hair on the back of your neck stand up. You are saying to yourself, with deep conviction, 'I don't care what anyone else says or does. I don't care what happens or what challenges I face. I don't care how hard it is. I am going to do this. I am going to change.' When you are at this point of serious, firm attention, your body sits up and pays attention. It knows the brain means business. And the body will now follow the direction of a firmly convinced prefrontal cortex."

PHYSICAL ACTIVITY AND ALZHEIMER'S DISEASE: SHRINKAGE HAPPENS

Dr. Cyrus Raji, M.D., Ph.D., is a bright, articulate, kindhearted doctor and researcher at the University of Pittsburgh's Department of Radiology, who oversaw and released some fascinating studies on the correlation between Alzheimer's disease, dementia, and physical activity. He has also become a good friend, as we both share a keen interest in the brain and longevity.

Dr. Raji comes to an interest in helping the brain stay healthier longer from personal experience. Cyrus's grandmother was a teacher, a brilliant woman who spoke five languages. But she was a smoker, suffered from a couple of strokes, and eventually succumbed to dementia and Alzheimer's disease. In her waning years of life her brain was reduced to that of a lost and confused child. Watching this bright light in his life grow dim, Cyrus has dedicated much of his career to doing what he can to stem the ugly tide of dementia and Alzheimer's in the world. As I mentioned in chapter 1, over five million Americans currently suffer from this disease, and because it doesn't just affect those who have Alzheimer's, but those who love them, the ripple effect of pain is tremendous.

Dr. Raji has been involved in brain imaging and Alzheimer's research for seven years, but for the last five years he's concentrated his research on how lifestyle factors can affect our brains positively or negatively. Using a special type of brain scan, researchers can measure the overall amount of volume in the brain as well as the volume in individual parts of the brain. The larger the volume in a brain, the healthier it is. When a brain is unhealthy or gets old, it will shrink, and the neurons will get smaller as well. But when a person has Alzheimer's disease, the neurons don't just shrink—they actually begin to die in the parts of the brain that are responsible for memory, organization, and personality.

Dr. Raji has been involved with a study that began in the 1980s, following 450 individuals over a period of twenty years, specifically to observe how lifestyle factors affected their brains as they aged. Some amount of brain shrinkage is normal in aging. Scientists call this process atrophy; the brain shrinks much the way a muscle will shrink when you don't use it. As the brain atrophies, senior moments happen with increased frequency.

However, in Alzheimer's disease, because the neurons are dying, there are actually fewer of them in the parts of the brain that help a person organize their day, remember things and people and places, and make up a large part of their personality.

One of the first studies that Dr. Raji worked on looked at how obesity affected brain volume in a group of normal people (individuals with no indication of brain deterioration issues). I mentioned this study in chapter 1, but it is worth repeating. He measured obesity using the body mass index (BMI), which is your weight divided by your height squared. A normal BMI is between 18.5 and 24.9; overweight is 25 to 29.5 (one hundred million Americans are in this category), and over 30 is considered obese (seventy-two million people in our country fall into this category). What Dr. Raji and his research team found was that the higher the obesity, the lower the brain volume and the higher the risk for Alzheimer's. Those who were overweight had some shrinkage. Those who were not overweight had no preshrinkage. This is the study upon which I base what I call the dinosaur syndrome in my PBS specials: the bigger the body, the smaller the brain. Not good. Dr. Raji's group repeated the study using seven hundred individuals with early stage Alzheimer's and found that obesity makes things worse. (They didn't use people in late stage Alzheimer's because people at this stage are thin since they are forgetting to feed themselves, and at this point, losing weight is not going to help their brain. It's too late.)

When Dr. Raji published this study, it received a lot of media attention. It was a "downer of a finding," Dr. Raji said in a recent conversation. This was the stimulus to find something positive that might help alter a bad brain trend. So he began to look at how lifestyle factors, particularly physical activity, might help the brain. He and his team looked at the most basic kind of simply physical activity anyone at any age can do: walking. He knew if they could prove that walking helped the brain, then it would follow that more exercise would do the same or perhaps even more good for brain volume.

"We looked at the effect of walking on 299 cognitively normal subjects," Cyrus shared. "We found that people who walked a mile a day or about twelve city blocks, six times a week, had increased brain volume over time in areas for memory and learning." Taking it further, he discovered that there would be a 50 percent reduction in the possibility of getting Alzheimer's over a thirteen-year test period. (Another way to say this is that Alzheimer's was cut down by a factor of two.)

In November of 2010, Dr. Raji looked at 127 people who had what we call mild cognitive impairment (lots of "senior moments") and were at a high risk for getting Alzheimer's or early stage Alzheimer's. "We looked at the effect walking had on their susceptible brains. In this study, we only had people walk about five miles a week, or three-quarters of a mile per day. The good news was that walking preserved their brain volume. It didn't increase their brain volume but it helped them to preserve what they had, without further brain shrinkage." This benefit even extended to those in the obese category. For anyone of any weight, walking stemmed the tide of brain atrophy.

Dr. Raji is often asked about other types of exercise for people who don't like walking. His reply is always, "Do what you like to do because you're more likely to participate in it more often. Being physically active improves blood flow to the brain. It delivers oxygen and other nutrients to your neurons."

JUST DO IT: KEEP AN EXERCISE STREAK

When software developer Brad Isaac asked Jerry Seinfeld, who in those days was still a touring comic, what his secret was, Seinfeld asked Isaac to pick up one of those wall calendars that had the entire year on a single page. To Seinfeld, becoming a better comedian meant writing every day, so for each day Jerry worked on his writing he would put a big red *X* in the box for that day. Pretty soon, there'd be a chain of red *X*s and not breaking the chain became its own motivation. Some people might think Andy McGill's dedication extreme, but it is often the hallmark of a successful person. Keep a good streak going.

There are moments when, caught up in the mental resistance that keeps us from getting started, we forget just how enjoyable the act of doing really is. When you've finally started and you're engaged in the work, you think, "Hey, I kind of like this." What I love about the Seinfeld calendar idea is that it lets you divert your stubbornness away from the "I don't wannas" and redirect to not wanting to mess up a good winning streak.

When I was in Sacramento recently for public television, the station manager used the same technique for exercise. He got on his treadmill for thirty minutes a day and marked an *X* on the calendar. So satisfying was that series of *X*s that he felt he had to keep his streak going. Then it became a habit. I urge you to try the "wall calendar and *X* system" as you begin your habit of daily exercise. This simple visual engages the brain in a way that motivates your body to get with the program. The chart appeals to your logical PFC; but the series of *X*s, signifying accomplishment, gives your limbic system a little rush of good feelings. Ka-ching! Your brain is buying in!

Bottom line? Find an exercise you like, whether it is walking around the block or hitting the gym or dancing with the stars (in your living room) and make up your mind, in firm intention to *just do it.* Your body and brain will thank you for decades and decades to come!

CHANGE YOUR AGE NOW: TWENTY BRAIN HABITS TO GET YOU MOVING FOR A LONG AND HEALTHY LIFE!

1. It is never too late to be the person you always wanted to be. Stopping a bad habit, like drinking too much, and adopting a new one, like exercising every day, is something anyone can do when they have truly made up their mind to change. Consider "doing an Andy McGill"—stop a bad habit today, and replace it with a new one right away.

2. Most "social drinkers" underestimate the amount they drink and the damage it is doing to their brains. Take a look at Andy's "before SPECT" scan again. If you know you are drinking too much, remember that excess alcohol is toxic to your brain. The "potholes" in an alcohol-soaked brain scan represent areas where your brain is not getting enough blood to function well. Determine to make your brain a "toxin-free zone" and get the blood pumping through it again!

3. Adopt Andy's attitude that daily exercise will ensure that one part of your day will be great! No matter what else happens, you can enjoy and feel energized in the special time you set aside to love yourself by investing in your health. Good mood endorphins are your immediate reward! Good health is the long-term payoff.

4. Start a streak and don't stop it! Try getting a calendar that you use only to keep track of your exercise. Make an *X* on it every day you exercise and set a goal to have five to seven *X*s on your calendar every week for the month. Reward yourself when you meet that goal! Then do it again, and again and again . . .

5. Your brain knows whether or not you mean business when you make a resolution. Set aside some time to let your brain and body know that your commitment to change is no joke by saying to yourself, with deep conviction, "I don't care what anyone else says or does. I don't care what happens or what challenges I face. I don't care how hard it is. I am going to do this. I am going to change." Write it down and reread your promise to yourself often.

6. Begin every morning by reminding yourself how you want to be and feel. Help create your own great day by doing this. Also remind yourself of how you don't want to feel and what steps you'll need to take to ensure you have a fabulous day—beginning with taking time out of your day for exercise.

7. Expect for it to take some time to push past the discomfort of adding a new routine, like daily physical activity, into your life. The brain likes the status quo, but it can be trained to change and upgrade itself. Take custody of your brain and body! Push through the unease, until exercise becomes a familiar and routine habit, like brushing your teeth.

8. Instead of reaching for sweets, fatty snacks, or alcohol when you are stressed, do something that will really work to lower anxiety and upset: Work up a sweat! Working out helps you manage stress by immediately lowering stress hormones, and it makes you more resistant to stress over time.

9. If you are suffering from mild to moderate depression or even a temporary low mood, remember that exercise can be as effective as an antidepressant without the negative side effects. In fact, the side effects are positive: You'll look better, have a better body, and increase your libido. Even if you take antidepressants for severe depression, exercise can boost the effects.

10. Besides longevity, looking good, and enjoying more energy, remind yourself that exercise is one of the best and proven preventions for dementia, cognitive decline, and Alzheimer's. If scientists could patent a pill with these kinds of results, they'd be very rich.

11. Everyone is motivated to get healthy for different reasons, but I have found that sharing the research finding that "as your weight goes up, your brain size goes down" is a powerful motivator for many people to get moving and get their weight under control.

12. Most people who do not exercise first thing in the morning, won't get it done. The day and its to-do lists invade us and provide excuses to miss our workout. Do your physical activity routine in the morning as part of your regular morning routine, and the habit will be easier to maintain.

13. Walking a mile six days a week is generally an easy and doable activity for most people and proven to help protect the brain. However, the best exercise for you is the one you will do! If swimming is your thing, by all means do that and enjoy it! If you like tennis, incorporate that into your weekly routine. Do whatever exercise floats your boat. The main thing is to simply do it, and do it almost every day, for consistency.

14. Consider dancing if you enjoy the music and the beat. It can keep you young at heart, can lift depression, keep you socially connected, and enhance your brain and your body.

15. Ever notice that you sleep better on days when you get physically active? And that you may have trouble getting to sleep and staying asleep when you've "couch potatoed" the day away? Engaging in exercise on a routine basis normalizes melatonin production

in the brain and helps give you a good night's worth of z's. Good sleep enhances your mood and decision making and also lowers your risk for obesity and depression.

16. Want to look younger? Those who do not exercise look older because they have more loose skin on their face and neck. Those who exercise tend to look years younger than those who do not.

17. Lifting weights or using your own weight as resistance can increase your muscle strength and stamina, and tighten up the muscles beneath your skin, giving you a whole body "lift." While watching TV, do some sit-ups and push-ups or leg lifts or squats. Keep a pair of handheld weights near the couch and pump some iron while you are watching the tube.

18. Get the lead out! Walking speed is a predictor of longevity, so try to put a little pep in your step. I tell people to walk as if they are late for a meeting.

19. Stretching and bending exercises such as yoga, Pilates, and tai chi help strengthen your core, support flexibility, reduce stress, and aid in balance, which can reduce your risk of falls.

20. To really burn calories, try burst training, which involves going all out for a few minutes followed by more moderate exercise. Run as fast as you can for a minute, and then walk fast for four minutes and repeat until thirty minutes or more is up.

4

JOSE

THEN WHAT? OPTIMIZE YOUR PREFRONTAL CORTEX TO FUEL CONSCIENTIOUSNESS AND MAKE BETTER, HEALTHIER DECISIONS

I don't tell lies, because I can look into the future and see that it is more trouble than it's worth.

—CHLOE, AGE 7

In large part, your behavior is driven by the actual, physical functioning of your brain. When your brain works right, you are more likely to act in thoughtful, conscientious ways that help you live longer. When your brain is troubled, you are much more likely to act in impulsive, careless, thoughtless ways that put you at risk for illness and early death. One of the smartest things you can do to increase the length and quality of your life is to optimize the physical functioning of your brain. Jose's story is a perfect example.

JOSE

In early 2010 a producer from the *Dr. Phil* show called and asked if I would help with a program on infidelity. They wanted me to evaluate and do SPECT scans on Jose, a compulsive cheater. When I first met Jose, he and his wife, Angela, were struggling with his infidelity, lies, and addiction to porn. As far as she knew, in their four-year relationship

Jose had cheated on her eight times. On the show, Dr. Phil replied to the eight incidences of infidelity by saying, "My father used to say for every rat you see there are fifty you don't."

They had been married three months when Angela found out Jose was cheating on her. She discovered that he had been with another girl when he asked her to marry him, when they were planning the wedding, and two days after they had Bella, their now three-year-old little girl.

"I was devastated and very angry," Angela said. "I gave my gun to my mom because I thought I was going to shoot him. After I took him back I found out he cheated on me with numerous girls. He is a chronic liar and he is very good at it. One of my friends told me that she saw Jose on a sex tape punching a girl in the face. She was passed out and pretty messed up. He likes rough sex and had tried to have rough sex with me. He tries to push to see how far he can go. He is a thrill seeker and needs constant stimulation. I got involved with Sexaholics Anonymous because I thought he had a problem, but instead he starting using it as an excuse. He would say, 'It's an addiction, I can't help it.' That's a bunch of crap. He doesn't think. He just does things and then afterward says he will find a way to deal with it."

Jose said, "I have always been the kind of guy who would just hook up when someone comes along. I was out of the home for five weeks before we decided to patch things up. Before we got married I never felt guilt. My father was a cheater. I am worried I have a sexual addiction because I have a need for something stimulating, such as affairs, fast cars, living on the edge. I lost my driver's license for getting four speeding tickets. In the last year, I have been faithful but had a problem with pornography."

On the show, Dr. Phil asked Jose, "If this is your proclivity, why not get a divorce and go do what you wanted to do?" Jose replied that it was not what he wanted. He wanted to be married, to have a family, and to raise his daughter. His father was a cheater, which had a negative effect on his family. He wanted to be a positive influence for his daughter.

When I saw Jose he had a number of issues besides the chronic infidelity. He was an adrenaline junkie who had a high need for speed along with excitement-seeking behavior. His brain SPECT scan showed three highly significant abnormalities:

1. Increased activity in a part of the front part of the brain called the anterior cingulate gyrus, which is the brain's gear shifter. Increased activity in this part of the brain is often associated with compulsive behavior, where the gear shifter becomes stuck on negative thoughts or negative behaviors. In addition to the cheating, Jose compulsively got tattoos. He was tattooed from head to toe. Even though Jose was a smart man, the tattoos had prevented him from getting work.

2. Decreased activity in another part of the front part of the brain, the prefrontal cortex (PFC). The PFC acts like the cop in your head and helps you stay on track toward your goals without going down the wrong path. The PFC is also thought of as the brain's brake and helps prevent us from saying or acting on the first thing that comes to mind. From Jose's scan and behavior, his PFC was in trouble.

3. A head-injury pattern. Jose's scan clearly showed evidence of brain trauma, with areas damaged in the front and back part of his brain.

Initially, I asked Jose if he ever had a brain injury.

He said no.

But as someone who has scanned the brains of tens of thousands of patients, I knew the pattern in Jose's brain was in part from a head injury, so I persisted.

Again he said no. I have heard this same story so often that it is a running joke at the Amen Clinics. People initially tell us they have not had significant head injuries. Then we see the obvious pattern of

Normal Surface Brain SPECT Scan

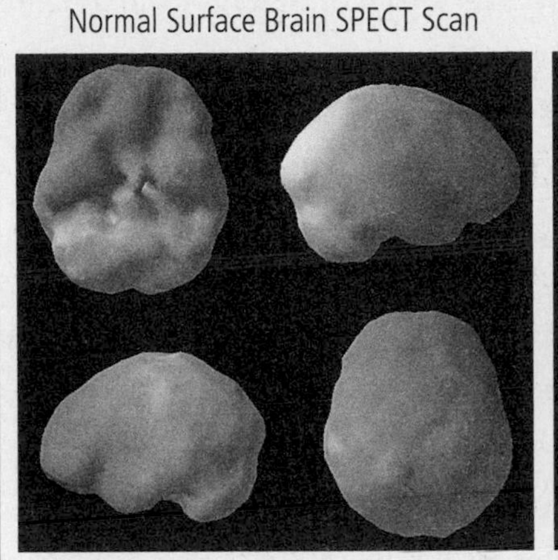

Full, even, symmetrical activity

Jose's Surface Brain SPECT Scan

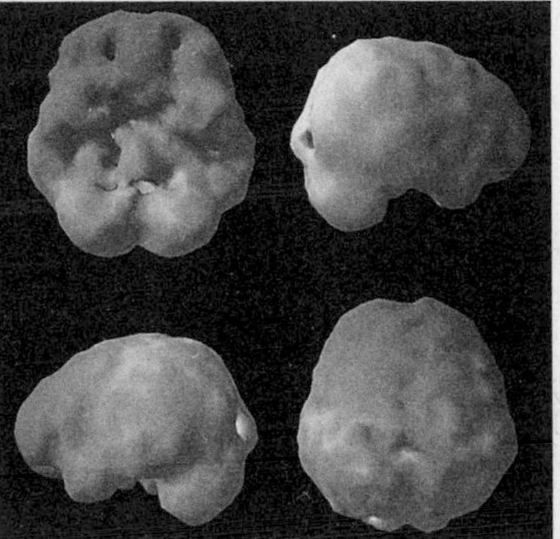

Decreased activity in the front (prefrontal cortex) and the back of the brain, consistent with a prior brain injury or injuries

Normal Active Brain SPECT Scan

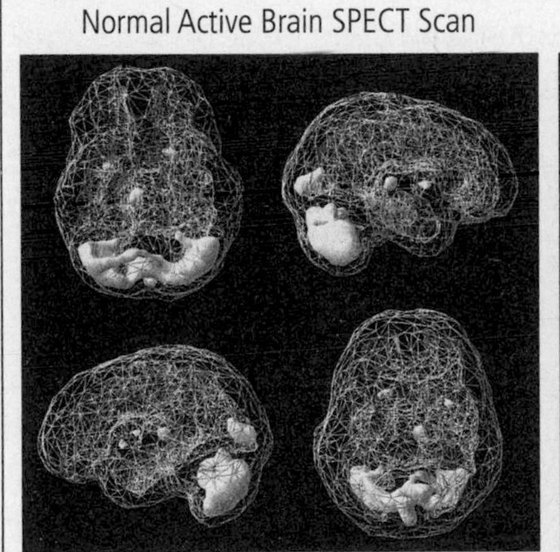

Highest activity at back of brain

Jose's Active Brain SPECT Scan

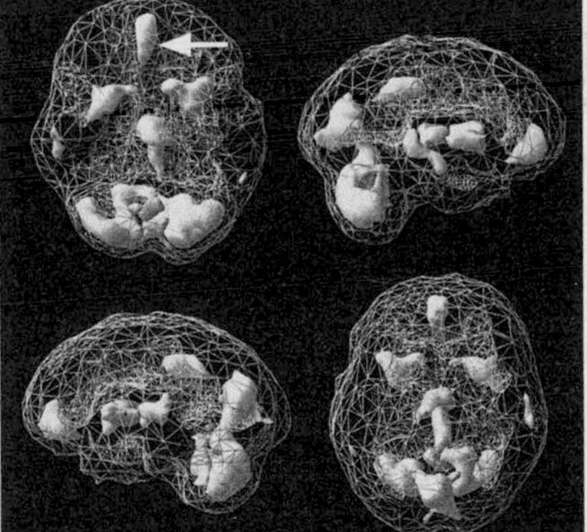

Increased anterior cingulate activity in front part of the brain consistent with trouble shifting attention (arrow)

a brain injury on their scans and persist in probing further. Eventually they will tell us things like, "I fell out of a second-story window" or "I fell down the stairs" or "I broke my car windshield with my head" in forgotten accidents. Or, as was the case with one of our NFL players, his car crashed through a mountain guard rail and fell 150 feet to a riverbed below, knocking him unconscious.

"Yes," Jose finally admitted. "I played football in high school." He then told me about a number of times he had concussions. Next, he volunteered that he was a bull rider and a mixed martial artist and had been hit hard in the head many times. And then, almost underneath his breath, he said, "And I am a head banger."

"Excuse me?" I responded.

With an embarrassed smile, Jose said, "I used to break things with my head. It was like a party trick. I could break cans and beer bottles with my forehead."

All of us have a running dialogue in our minds whenever we are talking to others. Psychiatrists are no different. When I heard Jose say that he used to break bottles with his head, I thought to myself, "This is not a sign of intelligent life." But I did not say it out loud, because I have a good PFC and a fairly strong internal brake.

But then Jose added, "When I got drunk I'd often put dents in doors and walls with my head. I can usually find the studs in the walls with my head."

At this, my own internal brake betrayed me, and I said out loud, "That is not a sign of intelligent life."

Jose agreed.

The day of the *Dr. Phil* show taping was emotional. Angela was angry and wanted Jose to change. She thought he could just will it to be so. Angela said, "Unless I see a complete change I am done."

I knew better. Even the best intentions are thwarted by an unhealthy brain.

On the show Jose said he was excited to see the results of the scans.

In his no-nonsense Texas drawl, Dr. Phil said, "It is odd to hear someone say they are excited to have brain damage. You think this gives you a pass. It is like, 'Hey, it's not my fault, my brain's not right.'"

Jose then said something very profound. "I am not thinking of this as an excuse, but I am hoping this might be a key to help change my behavior."

The show then took an interesting twist. Dr. Phil asked the audience whether they thought sexual addiction was a real biological phenomenon or just an excuse for bad behavior. The audience was of the opinion that it was just an excuse.

I understand why people feel this way, but from the brain scans I have seen and my years of experience in helping people unravel from addictions, I know there are strong brain issues at play. I have seen sexual addictions ruin people's lives and bring many addicts to the point of financial ruin and even suicide. I also believe that addictions, including sexual addictions, are going to get worse in our society as we are wearing out the brain's pleasure centers by the constant exposure to highly stimulating activities, such as video games, text messaging, sexting, Internet pornography, scary movies, and highly addictive foods like cinnamon rolls and double cheeseburgers.

There is an area deep in the brain called the nucleus accumbens that is responsive to the pleasure and motivation chemical dopamine. Think of the nucleus accumbens as one of the main pleasure levers in the brain. Whenever we feel pleasure, a little bit of dopamine has pressed on the lever. If the lever is pushed too hard, such as with drugs like cocaine, we can feel a rush of pleasure that causes us to lose control over our behavior, or if it is pushed too often, it becomes sensitized or numb and we need more and more pleasure in order to feel anything at all. Moreover, if you have low activity in the braking activity of the PFC, the nucleus accumbens can literally take control of your life, as in Jose's case. To live long, it is important to protect your pleasure centers and prefrontal cortex.

Even though it sounds odd, be careful with experiencing too much pleasure. I think one of the reasons actors and high-performance athletes have problems with depression and addiction is because their success can give them free access to anything they want at any given moment, and this often wears out their pleasure centers.

After the show, Jose and Angela agreed to see me for help. He was in enough pain that he was willing to follow my recommendations. Here was his prescription:

Stop drinking alcohol. Alcohol lowers Jose's PFC function and decreases his brain's braking power, which makes him less able to say no to his urges.

Get enough sleep to maintain healthy brain function. Getting less than six hours of sleep at night has been associated with lower overall blood flow to the brain, which leads to more bad decision making.

Clean up his diet. Only eat healthy food that serves his optimal brain function. Eat multiple times a day to keep his blood sugar stable. Low blood sugar results in more bad decision making.

Eliminate the caffeine and energy drinks that were a staple of his diet. Caffeine constricts blood flow to the brain. Anything that lowers or constricts blood flow to the brain increases bad decision making.

Add the following supplements to enhance his brain:

- Serotonin Mood Support to support healthy serotonin levels and calm his anterior cingulate gyrus and compulsive behaviors.
- Focus and Energy Optimizer to support healthy dopamine levels and boost his prefrontal cortex, focus, and impulse control.
- Brain and Memory Power Boost to help restore healthy brain function. This is the same supplement we used in our NFL brain rehabilitation study.
- High-quality fish oil.

Over the next seven months I regularly saw Jose, Angela, and their adorable daughter, Bella, to monitor their progress. In our sessions we discussed his nutrition, supplements, and strategies to control his urges, which were becoming less and less powerful.

I had Jose plant the question "Then what?" in his head to help boost his PFC by thinking about the future consequences of his behavior. It finally clicked when he heard the chorus of the Clay Walker song "Then

What?" Jose realized that if he didn't ask "Then what?" and make new, better choices he was going to be somebody who "ain't anybody anyone's gonna trust."

Things were going so well for Jose and Angela that they started to discuss having another child. They went to Hawaii on vacation to talk more about their future together. While there Jose saw people jumping off a sixty-foot cliff into the water below. His immediate reaction was that he wanted to do it too. Being a thrill seeker had been part of his life for a very long time. Some would say it was part of his DNA. As Jose hiked up the hillside, Angela rolled her eyes, thinking yet again to herself, "He is such a showoff." She had seen him do so many stupid things throughout their time together. Would it ever end?

But this time things were different. Very different.

When Jose got to the top of the cliff and looked down, something happened in his mind. He started to feel uncomfortable, even anxious. Even though he saw other people jumping off the cliff, he realized that because he could not clearly see the rocks jutting up in the water, it would be hard to avoid them. He thought to himself, "Then what? What if I land wrong? What if I get hurt? What if I am paralyzed? I have a wife and child and we want another child. Being paralyzed will not help any of us. Do I really need to do this?"

He stepped out of line to think about his next move. This level of thought, pausing to contemplate the consequences of a risky action, was new for Jose. After a minute or so, he decided not to jump. With a sense of freedom, he began walking down the hillside. Angela was stunned. She had never seen Jose do anything like that before. Maybe there was hope.

Shortly after his trip to Hawaii, we did a follow-up SPECT scan on Jose, which showed dramatic improvement from seven months earlier.

By working the treatment plan, Jose literally changed his brain and dramatically improved and likely extended his life. As I write this story it has been over a year and a half since I first met Jose, Angela, and

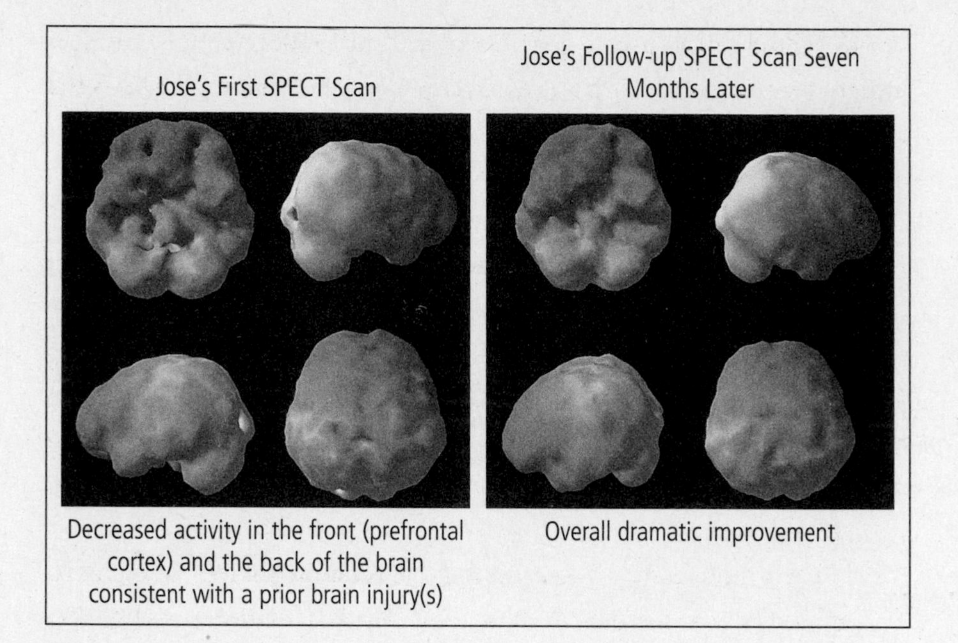

Decreased activity in the front (prefrontal cortex) and the back of the brain consistent with a prior brain injury(s)

Overall dramatic improvement

Bella. They remain happy, together, and hopeful about their future as an intact family. Angela no longer feels she needs to give her gun to her mother, and Jose has been faithful and is making decisions that will likely extend his life through better forethought.

CONSCIENTIOUSNESS AND LONGEVITY

I know many researchers are looking for the fountain of youth and hoping to find it in a new medication or natural supplement. I am rooting for these researchers to help us. But nothing will ever be as important to your longevity as the quality of the decisions you make throughout your life in regard to your health and your relationships. The quality of these decisions is a direct reflection of the physical health of your brain.

According to a remarkable longitudinal study, one of the main predictors of longevity is conscientiousness. The study was started in 1921 by Dr. Lewis Terman of Stanford University. He and his team evaluated

1,548 bright children who were born around 1910. Over the course of ninety years, researchers have discovered many fascinating findings that clearly point to healthy brain function and longevity. Here are some of their major findings:

- Hard work and accomplishment (usually associated with good brain function) are strong predictors of longevity.
- Those who were most disappointed with their achievements died the youngest.
- Being undependable and unsuccessful in careers (usually the sign of poor brain function) was associated with a whopping increase in mortality.
- Reaction to a loss with drinking, depression, anxiety, or catastrophizing was associated with early death (further causing poor brain function). On the other hand, those who, following a period of grief and adjustment (using brain healthy recovery skills), thrived after loss, got a "resiliency bonus" and lived an average of five years longer than average.
- An optimistic, carefree attitude encouraged people to underestimate risks and approach their health in a lackadaisical fashion, which decreased longevity. They died more often from accidents and avoidable deaths (behaviors associated with poor PFC function and subsequent poor planning). Some in the media have erroneously interpreted this study to mean that "pessimists live longer than optimists." This is not true; optimists of the hardworking, careful variety live longer lives than the average person. It is the "carefree" optimists who never worry, plan, or think about future consequences who do not live as long.
- Thoughtful planning and perseverance (usually associated with good brain function) were associated with longevity.
- Prudent, persistent achievers with stable families and social support lived longer (all signs of healthy brain function).

- People with habits, routines, and social networks that encouraged exercise did the best.
- Social relationships dramatically impact health. The group you associate with often determines the type of person you become. For people who want to improve their health, association with other healthy people is usually the strongest and most direct path to change.
- Moderate worry, meaning you care and think about the future, is an important part of staying healthy.

Clearly, this research and my own clinical experience have shown that some anxiety is good. People like Jose, who are risk takers and have low levels of anxiety, take unreasonable risks, which can lead to an early grave. Obviously, too much anxiety is bad. But not enough anxiety has been associated with more faulty decisions about health and safety.

People who are conscientious and "finish what they start" seem to have a reduced risk of developing Alzheimer's disease, according to a twelve-year study involving Catholic nuns and priests. The most self-disciplined individuals were found to be 89 percent less likely to develop Alzheimer's disease than their peers. Robert Wilson and colleagues from Rush University Medical Center in Chicago followed 997 healthy Catholic nuns, priests, and Christian brothers between 1994 and 2006. At the beginning of the study, the clergy completed a personality test to determine their level of conscientiousness. Based on answers to twelve questions such as "I am a productive person who always gets the job done," they received a score ranging from 0 to 48. On average, volunteers scored 34. Over the duration of the study, 176 of the 997 participants developed Alzheimer's disease. However, those with the highest score on the personality test—40 points or above—had an 89 percent lower chance of developing Alzheimer's than participants who received 28 points or lower. Dr. Wilson hypothesized that more conscientious individuals likely have more activity in their PFCs, and other researchers have confirmed this finding.

LOVE AND CARE FOR YOUR PFC

The PFC is larger in humans than in any other animal by far. It comprises:

- 30 percent of the human brain
- 11 percent of the chimpanzee brain
- 7 percent of your dog's brain (unless your dog is my dog, Tinkerbell, who won't stop barking at strangers; hers is likely 4 percent)
- 3 percent of your cat's brain (which is why they need nine lives)
- 1 percent of a mouse's brain

Neuroscientists call the PFC the executive part of the brain because it functions like the boss at work. It is the CEO inside your head. Comedian Dudley Moore once said, "The best car-safety device is a rearview mirror with a cop in it." Your PFC acts like the cop in your head that helps to prevent you from making bad decisions.

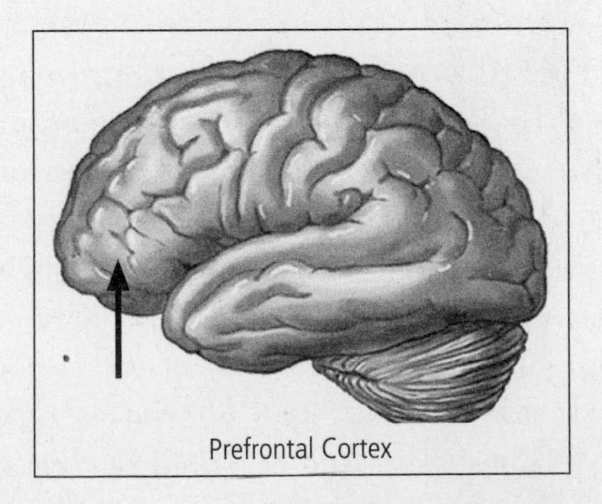

Prefrontal Cortex

It is like our own personal Jiminy Cricket. If it's not in good working order, ill-considered decisions can put you at risk for a miserable life, and even an early death.

The PFC is involved with:

- Forethought
- Judgment
- Impulse control
- Attention
- Organization
- Planning
- Empathy
- Insight
- Learning from mistakes

A healthy PFC helps you think about and plan your goals (e.g., "I want to live a long, healthy life"), and it keeps you on track for the long run.

Low activity in the PFC has been associated with:

- Lack of forethought
- Short attention span
- Impulsivity
- Procrastination
- Disorganization
- Poor judgment
- Lack of empathy
- Lack of insight
- Not learning from mistakes

The PFC is not fully developed until people are in their mid-twenties. Even though we think of eighteen-year-olds as adults, their brains are far from finished. Scientists are now learning what insurance companies have known for a long time. When do car insurance rates change? At age twenty-five. Why? Because that is when people display better driving judgment and are less likely to get into accidents and cost the insurance companies more money.

Below is a graph of activity in the prefrontal cortex across the lifespan. It is based on over six thousand scans we have done in our clinics. You can see that a child's PFC is very active, but over time the activity begins to settle down because unused connections are being pruned and brain cells are being wrapped with a white fatty substance called myelin.

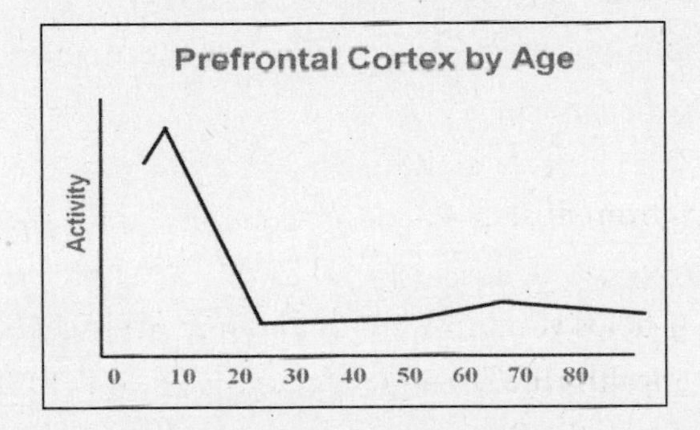

Myelin acts like insulation on copper wires and helps our brain cells work more efficiently. In fact, cells wrapped with myelin work ten to a hundred times faster than those without it. The prefrontal cortex is not fully myelinated, or efficient, until we are about twenty-five years old. Anything that disrupts myelin formation can actually delay or damage brain development. All of the following disrupt myelin formation:

- Smoking
- Drinking alcohol
- Drug use
- Brain trauma
- Poor diet
- Too much stress
- Not enough sleep

If we want our children to make better decisions for their lifetime, we need to do a much better job of taking care of their brains, since the PFC provides the horsepower for their decision-making skills for the rest of their lives.

Our research also shows a new burst of increased activity in the PFC after the age of fifty.

As I looked at this part of the curve, I began to think of the wisdom of age. Have you ever noticed that many "silly" things become less important as we age? We know the difference between "big stuff" and "small stuff"—and most of it really is "small stuff." We become more thoughtful and more able to focus on what really matters in life, which is why some people make better grandparents than they do parents.

Bill Cosby has a great routine in which he says that his children think their grandmother is the most wonderful person on the face of the earth. "I keep telling my children," he says, "that this is not the same woman that I grew up with. You are looking at an old person who is trying to get into heaven." That may be true, but more likely it has to do with the wisdom of age. If you take care of your brain, over time you are more likely to be wiser, because wisdom is a brain function based on intelligence born of many life experiences. Wisdom also helps us keep the Grim Reaper at a distance from our door.

Another way to think of the development of the PFC is to think about maturity. I think of maturity as not making the same mistakes over and over. Mature people have a more thoughtful approach to their lives. As the PFC becomes myelinated, people act with more forethought and are likely to make fewer mistakes.

THE DINOSAUR SYNDROME REVISITED: BIG BODY, LITTLE BRAIN, BECOME EXTINCT

I, along with my colleagues Kristen Willeumier, Ph.D. (a director of research) and Derek Taylor (a data analyst), published an important study in the prestigious N*ature Publishing Group Journal,* "Obesity," which demonstrated that as weight went up in a healthy group of people, the function of their PFC went down in highly significant ways.

Over the last decade there is growing evidence that shows the harmful effects of too much fat on your body. In a study of 1,428 Japanese men, researchers found significant decreases in brain size in the PFC and temporal lobes (learning and memory). Nora Volkow, director of the National Institute on Drug Abuse, and colleagues found that in healthy adults a high BMI (body mass index) was inversely correlated with activity in the PFC. Elevated BMI has also been associated with myelin abnormalities in the PFC of healthy normal and elderly adults.

The goal of our study was to test the hypothesis that an elevated BMI is associated with lower blood flow to the PFC in a healthy group of people on brain SPECT imaging. To that end, we compared our group of "healthy" subjects who had a high BMI with people in our "healthy" group of normal weight. The results of our study were very clear. The high BMI group had statistically significant lower activity in the PFC compared with the normal group.

Obesity is becoming a worldwide epidemic and is a risk factor for many chronic conditions, including cardiovascular disease, depression, and neurodegenerative diseases like Parkinson's and Alzheimer's. It has been recently found to be worse for your liver than alcoholism.

We were not able to determine if problems in the PFC led to increased impulsivity and subsequent obesity or if being overweight or obese directly caused brain changes. Both scenarios may be true. The fact that we used a healthy-brain group and specifically excluded ADHD or other behavioral disorders argues against the premorbid hypothesis, but other studies have shown an association between ADHD and obesity. Still other authors report that fat tissue directly increases inflammatory chemicals, which likely have a negative effect on brain structure and function.

One of the major problems with being overweight or obese is that there is evidence that it damages your PFC, which as we have seen is the major decision-making part of the brain. So if you do not get your weight under control, it will become harder and harder to use your own good judgment over time to get and stay healthy. *Now* is the time to start enhancing your health and longevity, not at some arbitrary point in the future, which most likely will never come.

ADHD, PFC PROBLEMS, AND EARLY DEATH

ADHD is associated with low activity in the PFC. Initially, ADHD was thought of as a childhood disorder that most kids outgrew by the time they turned twelve or thirteen. The hallmark symptoms of ADHD are short attention span, being easily distracted, disorganization, hyperactivity (trouble sitting still), and poor impulse control. People with ADHD, like Jose, often exhibit excitement-seeking or conflict-seeking behavior; they also tend to have trouble with time (they are often late and turn in assignments at the last minute). Over the last three decades it has become clear that many ADHD children continue to have debilitating symptoms for the rest of their lives. They tend to outgrow the physical hyperactivity but not the problems with disorganization, inattention, distractibility, and impulse control. Untreated ADHD has been associated with a higher incidence of:

- Drug and alcohol abuse (impulsivity and to calm feelings of hyperactivity)
- Relationship problems (impulsivity and conflict seeking)
- School failure (attentional problems and impulsivity)
- Job-related problems (problems with time, attention, and impulse control)
- Medical problems (associated with chronic stress, plus more head trauma with the excitement-seeking behavior)

- Obesity (lack of impulse control)
- Depression (chronic failure)
- Lack of conscientiousness (all of the above)

In the book I wrote with noted neurologist Rod Shankle, *Preventing Alzheimer's*, we argued that ADHD is likely associated with Alzheimer's disease because of its connection with many of the illnesses that put people at risk for it, such as alcohol abuse, obesity, depression, and head trauma. This is very important, because when ADHD goes untreated, a person will not be able to control his or her impulses, setting him up for significant health problems, poor decisions, and earlier death. If you or someone you love has symptoms of ADHD, it is important to be treated. Natural ways to treat ADHD, in my experience, include intense aerobic exercise, a very healthy diet, a multivitamin, fish oil, and supplements (such as green tea, rhodiola, L-tyrosine) or medication (such as Ritalin or Adderall) to enhance PFC function.

Once you realize the absolutely critical role of the PFC to longevity, you then need to do everything possible to protect it and rehabilitate it if necessary.

BOOST YOUR PFC TO REIN IN YOUR INNER CHILD AND BOOST CONSCIENTIOUSNESS

All of the information in this book is designed to help you win the war in your head between the adult, thoughtful part of your brain (the PFC), which knows what you should do, and your pleasure centers, which are run by a spoiled, demanding inner child who always wants what he wants whenever he wants it.

Your brain's pleasure centers are always looking for a good time.

- They want to jump off a cliff.
- They love going fast on a motorcycle in the rain.

- They crave the ice cream.
- They want the double cheeseburgers.
- They will stand in line for the fresh cinnamon rolls.
- They focus on having the second piece of cake.

Left unchecked, your inner child is often whispering to you like a naughty little friend:

"Do it now . . ."
"It's okay . . ."
"We deserve it . . ."
"Come on, let's have some fun . . ."
"You're so uptight . . ."
"Live a little . . ."
"We already had one bowl of ice cream. Just one more won't hurt . . ."
"We'll behave better tomorrow. I promise . . ."

Without adult supervision, your inner child lives only in the moment and he can ruin your life. I have a friend who shared that her daughter-in-law got sick with the flu and had to remain in bed upstairs. Her four-year-old son decided to "take over" the house while his mother was otherwise occupied with a pounding headache, fever, and vomiting. When his father came home, there was ice cream melting in puddles on the kitchen countertop; the pots and pans were arranged in a pyramid formation in the middle of the floor; cartoons were blaring on TV at full volume; and clothes, toys, and blankets (made into tents and forts) were strewn everywhere. Absolute anarchy and chaos. This is a great visual of what happens to your life when your PFC is not functioning: Your inner child takes over while your inner adult is napping. The resulting mess is something to behold.

To balance your pleasure centers, and tame your inner child, the PFC helps you think about what you do before you do it. It thinks about your future, not just about what you want in the moment. Instead of thinking about the chocolate cake, it is the rational voice in your head that helps you:

- Avoid having a big belly.
- Remember that "food is medicine" and that you'll be in a sugar-induced, cranky, sleepy mood an hour after eating that cake.
- Remind your inner child of delicious but healthier alternatives that will both taste good and nourish your body.
- Be concerned about your bulging medical bills.
- Say no and mean it.

When your PFC is strong, it reins in your inner child, so that you can have a fun, passionate, meaningful life but in a thoughtful, measured, conscientious way. To live a long healthy life, it is critical to strengthen your PFC and put your inner child into time-out whenever he acts up.

It is also critical to watch your internal dialogue and be a good parent to yourself. I have taught parenting classes for many years and the two words that embody good parenting, even for your inner child, are *firm* and *kind*. When you make a mistake with food or with your health, look for ways to learn from your mistakes but in a loving way.

CAN YOU CHANGE YOUR LEVEL OF CONSCIENTIOUSNESS?

Changing one's personality traits is never easy. They are thought to be enduring patterns that ultimately come from stable patterns of brain function. But in the Terman study, researchers found that people can indeed increase or decrease their conscientiousness over time. Jose was able to do this, and I have witnessed it in myself. As I have learned more and more about brain function and developed brain envy, I have personally developed better habits and my behavior has been more consistent. I feel much more in control of my own behavior than I did even four or five years ago. I have seen others' conscientiousness deteriorate after a head injury, binge drinking or drug use, being exposed to an environmental toxin, or at the onset of developing dementia.

Before discussing how to boost your level of conscientiousness, let's first define what it is. Conscientiousness concerns the way we manage our impulses. Impulses are not inherently good or bad. It is what we do with them that makes them that way. Sometimes we need to make a snap decision and cannot think about it over and over. Other times we want to be spontaneous and fun, especially when we are relaxing. But when it becomes a way of life, it can take a seriously negative toll on your health. Giving in to immediate desires, like the doughnuts, often produces immediate rewards but undesirable long-term consequences. Impulsive behavior can lead to being fired from your job, divorce, drug or alcohol abuse, jail, or obesity, all of which have a negative impact on your health. Acting impulsively often brings regret because you failed to entertain all of your options. The accomplishments of an impulsive person are often smaller, more diffuse, and less consistent.

A hallmark of intelligence and what separates us from other animals is our ability to think about the consequences of our behavior before acting on an impulse. It is the internal dialogue that accompanies "Then what?" Effective decisions usually involve forethought in relation to your goals, organizing, and planning, which helps you not only live in the moment but to continue ten or even fifty years from now. "Being prudent" is another label for conscientiousness. It means being wise and cautious. If you are conscientious, you are more likely to avoid troubled situations and be perceived as intelligent and reliable by others. If you go overboard, of course, others will think you are a compulsive perfectionist or a workaholic.

SIX FACETS OF CONSCIENTIOUSNESS

1. *True confidence.* You have a true feeling of being self-efficacious. You know you can get things done.

2. *Organized, but not compulsive.* Keep an orderly home or office and keep lists and make plans.
3. *A high sense of duty.* You have a strong sense of moral obligation.
4. *Achievement oriented.* Drive to be successful at whatever you do and have a strong sense of direction.
5. *Persistence.* You have the ability to stay on track despite the obstacles that might come your way.
6. *Thoughtfulness.* You are disposed to think through possibilities and the consequences of your behavior before acting.

Here are steps to optimize both your PFC and level of conscientiousness to boost the control you have over your life.

1. **"Then What?"** Always carry this question with you. Think about the consequences of your behavior before you act.

2. **Protect Your Brain from Injury or Toxins** This should be obvious by now.

3. **Get Eight Hours of Sleep** Less sleep equals lower overall blood flow to the PFC and more bad decisions.

4. **Keep Your Blood Sugar Balanced Throughout the Day** Research studies say that low blood sugar levels are associated with lower overall blood flow to the brain, poor impulse control, irritability, and more unfortunate decisions. Have frequent smaller meals throughout the day that each have at least some protein.

5. **Optimize Your Omega-3 Fatty Acid Levels by Eating More Fish or Taking Fish Oil** Low levels of omega-3 fatty acids have also been associated with ADHD, depression, Alzheimer's disease, and obesity.

6. **Keep a "One-Page Miracle"** On one piece of paper write down the specific goals you have for your life, including for your relationships, your work, your money, and your health. Then ask yourself every day, "Is my behavior today getting me what I want?" I call this exercise the One-Page Miracle, because it makes such a dramatic difference in the lives of those who practice it. Your mind is powerful and it makes happen what it sees. Focus and meditate on what you want.

7. **Practice Using Your PFC** Self-control is like a muscle. The more you use it, the stronger it gets. This is why good parenting is essential to helping children develop self-control. If we gave in to our eight-year-old every time she wanted something or threw a temper tantrum, we would raise a spoiled, demanding child. By saying no and not giving in to tantrums, we teach her to be able to say no to herself. To develop your PFC you need to do the same thing for yourself, practice saying no to the things that are not good for you, and over time you will find it easier to do.

8. **Balance Your Brain Chemistry** Illnesses such as ADD, anxiety, and depression decrease self-control. Getting help for these problems is essential to being in control of your life.

Trying to use willpower to control your behavior when your sleep or brain chemistry is off, or when your omega-3 fatty acids or blood sugar levels are low, is nearly impossible.

BE THE BOSS OF YOUR LIFE AND YOUR LONGEVITY

When I walk down the street and see people who are not healthy, I often say to myself, "That person has made many, many bad decisions." It frustrates me because I know with the right education and right environment they would be healthier and happier. When I see someone who is healthy I think, "That person has made many, many good decisions." It is the quality of the decisions you make that helps you live a long time as a healthy human being or kills you early. By applying the principles in this book you can boost your PFC and have much better control of your health and your destiny. You can be the boss of your life, instead of allowing your cravings or the food companies to take your life early. A little forethought and an appropriate level of anxiety are all that is required. When confronted with choices between a spinach salad or a double cheeseburger, between going to a late-night party or getting a good night's sleep, between going for a hike or jumping off cliffs into the water below, can you step back and ask yourself which choice is really in your best interest? Does the choice you favor make you better, stronger, healthier, more passionate for your life? Or does it steal from your life? Choose to be in charge. Choose to be the CEO of a long, healthy, vibrant, meaningful life.

CHANGE YOUR AGE NOW: TWENTY TIPS TO HELP YOU MAKE BRAIN HEALTHY DECISIONS TODAY

1. When your brain is troubled, you are much more likely to act in impulsive, careless ways that put you at risk for illness and early death. When your brain works right, you are more likely to act in thoughtful, conscientious ways that help you live longer. Prioritize your brain health and better behavior will follow.

2. Nothing is more important to your longevity than the quality of the decisions you make in your life. And the quality of your decisions is a direct reflection of the physical health of your brain. Taking time to look at and upsize your brain health may be the most important decision you make to live strong and long.

3. Decreased activity in the PFC has been associated with lack of forethought and poor judgment. When it does not get enough blood flow, you don't have a good working brake on your impulses. Increasing blood flow to this area with brain healthy habits, along with supplements, such as green tea and rhodiola, will help a person make better decisions, leading to a longer and happier life.

4. Head injuries, even minor concussions from the past, show up on SPECT scans and may affect your behavior and feelings years later. We often have to ask people many times, "Have you ever had any sort of injury to your head?" before they recall the incident that hurt their brain. Recognizing and rehabilitating these injuries will dramatically increase the quality of all of your decisions.

5. Addictions, including sexual addictions, are made worse when we literally "wear out" the brain's pleasure centers by the constant exposure to highly stimulating activities, such as video games, text messaging, sexting, Internet pornography, and scary movies. Take inventory of the adrenaline-producing activities in your life: Eliminate the unhealthy ones, and take breaks from those activities (even good ones) that are becoming compulsive.

6. An overly optimistic, worry-free attitude (without forethought and planning, which are PFC activities) leads people to underestimate risks and approach them in lackadaisical fashion, decreasing longevity. Be optimistic, as this is good for longevity; but balance this trait with a healthy level of anxiety and careful thinking.

7. People who were persistent achievers with stable families, habits, and routines did the best in longevity studies. Consider your daily and weekly routines: Can you make them more brain friendly? For

example, can you walk somewhere that you always drive to? Can you exchange an hour of TV for playing brain games?

8. To make better decisions, make sure to optimize the blood flow to your brain by stabilizing your blood sugar (make sure you eat healthy and often), getting good sleep, limiting alcohol and caffeine, and eliminating nicotine.

9. If we want our children to make better decisions for their lifetime, we need to do a much better job of taking care of their brains. The brain is not fully efficient until we are twenty-five years old. To avoid disrupting early brain development, help young people avoid smoking, substance abuse, brain trauma, a lousy diet, stress, and poor sleep.

10. Our research shows a new burst of increased activity in the PFC after the age of fifty. We become more thoughtful and more able to focus on what really matters in life. Have a few wise friends over age fifty who can give you valuable insight when making decisions.

11. When fighting addictions or everyday temptations of any kind, always keep this question in mind: "Then what?" Whenever you think about doing or saying something, ask yourself about the consequences of your behavior. This question can serve as a caution or stop sign to a brain that is about to take you down a bad path.

12. A hallmark of intelligence and what separates us from other animals is our ability to think about the consequences of our behavior before acting on an impulse. It is the internal dialogue that accompanies "Then what?"

13. A high percentage of people who struggle with addictions also have ADD or ADHD. When left untreated, a person has less ability to control his or her impulses, setting him up for significant health problems, poor decisions, and early death. Natural ways to treat ADHD include intense aerobic exercise, a very healthy diet, a multiple vitamin, fish oil, and supplements (such as green tea, rhodiola, L-tyrosine) or medication (such as Ritalin or Adderall) to enhance PFC function.

14. Being overweight is damaging to your PFC and can have a negative impact on the decision-making part of the brain. Getting your weight under control, starting now, will help you enhance your health and longevity.

15. Practicing self-control is a good exercise to strengthen your PFC. The more you use it the stronger it gets. To develop your PFC you need to practice saying no to the things that are not good for you, and over time you will find it easier to do.

16. Low levels of omega-3 fatty acids have also been associated with ADHD, depression, Alzheimer's disease, and obesity—all brain issues that lead to poor decision making. You can optimize your omega-3 fatty acid levels by eating more fish or taking fish oil.

17. Keep a One-Page Miracle. On one piece of paper write down the specific goals you have for all the main areas of your life. Then ask yourself every day, "Is my behavior getting me what I want?" This simple but profound activity can be of tremendous help in encouraging better daily choices that add up to a better life.

18. On occasion, we all need to rein in our inner child who wants to eat junk, avoid sleep, or "play with fire." But it is important to be a

good parent to yourself, which means being firm and kind. When you make a mistake with food or with your health, look for ways to learn from your mistakes but in a loving way. Emotionally healthy people deal with mistakes by acknowledging them, learning from them, and moving on from them as soon as possible.

19. Hard work and accomplishment (usually associated with good brain function) are strong predictors of longevity.

20. The quality and length of your life are direct reflections of the quality of the decisions you have made.

5

JIM

BOOST YOUR BRAIN'S LIFE SPAN, SPEED, AND MEMORY

You know you've got to exercise your brain just like your muscles.

—Will Rogers

Once you optimize the physical functioning of the brain, it is then critical to keep it strong. Mental workouts and lifelong learning strategies are essential tools to keep your brain young, agile, and adaptable. One of my friends spends his life teaching people just how to do this. Jim's real last name is Kwik. Which is uncanny, because today he helps others "speed up" their brain's processing abilities. His list of clients is impressive, and he flies all over the globe teaching corporations, executives, doctors, lawyers, and students the secrets of how to improve brain function that he once stumbled upon in his lowest moments.

"MY INSPIRATION WAS MY DESPERATION"

"My inspiration was my desperation," Jim told me. "I was a young guy, a freshman attending college in New York. And I was struggling. The jump from high school to college was especially tough for me." Jim was not prepared for the amount of material he was expected to read, assimilate, and regurgitate for his courses. "I worked harder than most of my friends, and this was tough on my self-esteem, as it is for anyone who struggles in school." At school, like many first-year college students I've treated, Jim was in a state of overwhelming stress, the likes of which he'd never known before. He was juggling midterm exams, labs, and writing papers, on top of trying to keep up. He went without sleep and food, studying and reading day and night. One day he simply passed out. "Two days later I woke up dazed and confused, and in a hospital. I had been trying so hard to learn and read everything, until I literally put my brain and body into a state of exhaustion. I was trying to drink water out of a fire hose and was barely surviving."

Jim thought to himself, "There has to be a better way." His answer walked in the door, right at that moment, carried by a nurse. She held a hot mug of tea and handed it to Jim. On the cup was a quote by Einstein: "Problems cannot be solved by the same level of thinking that created them." Looking at that quote, Jim got chills down his spine.

Jim was looking at the problem—his need to absorb information—and trying to solve it the same way it was created, by pushing himself harder to absorb more and more. He had been working harder. Now it was time to work smarter. It was in that life-altering moment, when he was vulnerable and open to another way, that he realized, *School is great for teaching us* what *to learn. Where it fails is teaching us* how *to learn.*

Jim began asking himself questions: "How can people learn how to remember things more easily? How do they use their brain to improve their focus? How can I train my brain to better handle the hordes of information coming at me? Is there an easier way than rote learning?"

INFORMATION OVERLOAD AND THE SECRETS TO BECOMING A LIFELONG LEARNER

Our modern minds have to handle more information than ever before in history. For example, information is so fast-paced that it doubles now every two years. We have a half million words in the English language, five times as many words as in the time of Shakespeare. Someone who goes to college for a four-year degree may discover that by their third year of study, much of what they've learned is already outdated. *There's more information in one issue of a* New York Times *than a person in the eighteenth century would have been required to digest in his whole lifetime.* If you look at the top in-demand jobs for 2011, most of them didn't even exist in 2004. That's *a lot* of change, a lot of information.

Jim's pain created questions, which in turn led to life-changing answers. He began studying every book he could find on how to quicken the brain's ability to learn. What Jim learned radically changed his performance in school. He wasn't working as hard as before, but he got better grades with less effort. He began to realize that if he could do this, anybody could. "There's no such thing as people who have better memories than others," he now tells his students. "There is a trained memory and an untrained memory."

Science verifies what Jim teaches. With a simple plan you can dramatically improve your brain's ability to think and remember. But to do this you have to use your brain on a regular basis. Here are some immediate tips to implement in your life.

Dedicate yourself to reading something that interests you for thirty minutes a day. The brain is like a muscle: you need to use it or you will lose your ability to use it. The biggest mental declines happen after we complete our formal schooling and after retirement. Why? Because we are not pushing ourselves to continue to learn, grow, and stretch our neurons. Reading helps the learning continue. People who are in a job that does not require continual learning are at greater risk for Alzheimer's disease.

Turn your car into a "university on wheels." Listening to audiobooks is another way to keep your mind active and sharp. When I walk, I love listening to the latest audiobook I've downloaded to my smartphone, which actually makes me smarter.

Journal every day. You'd be amazed at how many of the great men and women of history kept a journal. Journaling can take many forms, from the conventional pen-and-notebook-style journals to blogging or simply posting meaningful quotes, thoughts, and experiences on social media sites.

Stay childlike when it comes to learning. Jim says, "My ninety-five-year-old grandmother is one of the youngest people I know." It is because she has kept her childlike curiosity intact. Did you know that preschoolers ask between three hundred and four hundred questions a day? Not only should we never stop asking questions, but we should be actively curious. Ask yourself, "What if?" and then seek out the answers.

Keep your emotional state primed for learning. All learning is dependent upon your emotional state. A brain that is emotionally balanced is ready and primed for learning. When we are bored, cranky, and tired, it doesn't matter how interesting the teacher might be, we are not going to learn anything new. If we are depressed or stressed or obsessed, all our brain energy is being used to try to prop up our emotions for survival; very little is left for new learning. *Use your body to train your mind.* The research is conclusive: Exercise helps the brain learn better. We are all pretty familiar with the mind–body connection now. But fewer realize the body–mind connection, or how the body helps stimulate different parts of the brain. For example, Jim teaches people to use their finger or a pen to follow the words on a page. This one simple act of using the body to help the mind read will give you 25 percent increase in speed and focus.

Create a positive learning environment. All learning is "state dependent." "We learn best," Jim points out, "when we can get in that state known as flow, when we are alert and relaxed at the same time." Other scientists call this a concert state: a calm but focused mind, such as when you attend a symphony. Stress is the enemy of learning. We tend to freeze up when stressed and we cannot take in new information or process it well in that mood state. Studies have shown that students score higher on tests of recall and memory after they have been shown a funny movie clip. The laughter relaxes their brain so that it is optimized and open to absorb new information.

It pays to proactively create a positive environment for learning. Rooms that are either too boring or too busy can distract from learning. Good lighting is key. Some people learn better when there is beautiful art or music around. Baroque music that plays at sixty beats per minute has been shown to help learning. Smells actually can anchor learning. Jim has his students learn something new while wearing a particular scent or smelling an essential oil. Later when they take the test or need to recall the information, he asks them to wear the same scent or take a whiff of the essential oil and they'll recall the information better. The olfactory senses are very tied to memory. We've all had experiences of catching a certain scent, like bread baking, and being drawn back to a warm memory.

MEMORY TIPS

Mental practice is to the mind what physical exercise is to the body. One way to enhance your brain is to boost your memory skills. Brain cells in an area of the temporal lobes, called the hippocampus, are responsive to training. These are some of the first cells to die in Alzheimer's disease, so working to keep them young is critical to lifelong brain health.

Jim teaches his clients how to boost their memory abilities. We are all emotionally connected to our names, so when someone remembers our name and addresses us by it, it makes us feel special. Here are some tips that Jim shared, which you can use to help remember people's names. Note that you can also use these strategies to help you remember other items. The acronym SUAVE spells out the key steps.

S: Say the person's name. When someone tells you his name, repeat it in a natural way. For example, someone tells you, "I'm Joshua." Repeat his name in a sentence: "Joshua, nice to meet you."

U: Use the person's name. In a natural way, use this person's name again as the conversation goes on. "Hey, Joshua, can I get you a cup of coffee?" (Remember to use it, but don't abuse it. If you use his name in every other sentence you'll start to sound like a salesman about to make a pitch.)

A: Ask questions. This is especially good with unusual names. Ask the person, "How do you spell that?" Or comment, "That's a beautiful and unusual name. Do you know its origin?" Or "What does it mean?"

V: Visualize the person's name. Create a funny or unique or crazy image in your mind. For example, if a person's name is Mark, imagine you putting a check*mark* on his forehead. If his name is Michael, think of him grabbing a "*mic*rophone" and jumping up on a table to sing karaoke. For Alexis, picture a woman driving a Lexus. The wilder the image, the better it will stick in your brain. People sometimes ask Jim how he remembers so well and he responds, "How can I forget it? You should see the crazy picture I created in my mind!"

E: End every conversation with using their name before you say good-bye. "It was great visiting with you, Bob," you might say as you mentally visualize him "*bob*bing" for apples just one more time.

Decide to focus on remembering people's names for twenty-one days to embed the habit. "I'd even make up names for people when I walked into a store," says Jim, "just to see if I could remember their 'made up' names when I left the store."

SPEED READING

Since reading can help work out your brain, let's enhance it with speed reading. Jim is an expert on teaching people this skill, so I asked him to share some secrets with you. He says, "Reading faster is a skill that anyone can learn. But to become a speed reader, it means to let go of or 'unlearn' what has been a comfortable, familiar, albeit slow habit. And the hard part is that sometimes you have to slow down during relearning before you can speed up." Jim told me that as a boy he mastered the art of typing with two fingers when he'd stay at his grandparents' home. "They were wonderful loving people, but they had no toys," he explained. So he kept himself entertained on their old typewriter, by teaching himself to type like crazy with two fingers.

And then he took a required typing class at school.

The teacher asked him to let go of his two-finger method and use all ten of his fingers instead. What do you think happened to his typing speed at first? Yes, it slowed way down. But eventually, when he mastered the art of ten-finger, or touch, typing, he was able to type faster than ever. "Reading is much the same way," Jim explains. "Most people are reading with two fingers, so to speak."

Here are some insights and tips to help quicken your reading speed.

- Though you may think that people who read faster comprehend less, the opposite is true. Here's why. People who read slowly read One. Word. At. A. Time. They are reading so slowly that they are boring themselves. Their mind begins dashing around the environment looking for something more interesting to hold its attention. They can't focus on the content of what they are reading. Faster readers actually have better comprehension because they can focus more easily: Basically, the information is hitting their brain at a more interesting speed.
- Another common issue that slows down reading is "subvocalization," which means that some people say aloud every word they are reading in their head. Using this method, you can only read as fast as you can talk, which is 200–500 words per minute. We think much faster than we talk, so by getting rid of the subvocalization habit, with specialized training, you can begin to read closer to the speed of thought than the speed of talking.
- Regression or rereading slows down reading. This habit is like someone having control of a DVD and rewinding it a little bit

about every thirty seconds. Breaking this two-finger habit helps people to read faster.

- Using a finger, a pen, or a computer mouse to follow the words, as if you are invisibly underlining the sentences, will increase your reading speed 25–50 percent across the board. The reason? Your eyes are attracted to motion, and this increases focus. Also, in the way that the senses of taste and smell are linked, so are touch and sight. There is a touch–sight connection in the brain that when activated increases speed and comprehension.
- If you are right-handed, try using your left hand to follow the words as you read. This activates more of your whole brain. Most people are "left brain readers," and they find when they use this method that it engages their right brain. One of Jim's clients said that he reread Ernest Hemingway's classic *The Old Man and the Sea* using this method, "only this time around, it was like I could actually feel the sand on my feet and hear the ocean waves. The only thing I didn't like was the smell of the fish."
- Take notes as you read. By taking notes as you read, your comprehension will shoot up. If you share or relate what you read, even pretend to "teach it" to someone else, your retention will be even higher.

MIND WORKOUTS FOR EVERYONE: 24/7 BRAIN GYM TRAINING FOR BOOSTING YOUR BRAIN

THE ELEPHANT AND THE RIDER: SYNCING TWO FORCES

In *The Happiness Hypothesis*, author and philosopher Jonathan Haidt uses the metaphor of an elephant and a rider to help us visualize two strong forces in the brain. The prefrontal cortex, or PFC, is much like the rider and involves the thinking, logic center that we assume (or like to assume) is in control of our lives. The elephant is what I would call the limbic system and represents our emotions, which are automatic responses to outward triggers based on stored memories. As long as the elephant wants to go where the rider directs him, things work fine. But when the elephant "truly, deeply, madly" wants to go somewhere that the rider prefers him not go, who is going to win that tug of war? Most bets are on the elephant.

How do we integrate, then, our own rider and elephant so that our PFC and limbic brains, our goals and our desires, our thoughts and behaviors, get more in sync? One way to do that is through continual, goal-directed, brain-training techniques. This is one of the reasons we developed the brain-training modules on our website at www.theamensolution.com. I like to call it our 24/7 Brain Gym, because you can log on and work out your brain at any time.

The process starts with a long assessment to help individualize your program. Based on how you score, you're given a personalized set of exercises to boost your weak areas and strengthen the ones that are already working well. The development of this part of our site flowed out of years of gathering information from thousands of people, using the latest research on how to optimize brain function. This allowed us to create a program that helps the brain operate better as a whole system, ultimately helping your behaviors and your beliefs work together. In other words, it helps the rider and the elephant stop the tug-of-war and work together in cooperation.

Savannah DeVarney, one of our site developers, says, "What we have found is that one brain function precedes the other. Our internal state drives our external state, or how we behave in the world. Every fifth of a second you have an emotion that becomes a feeling, which may turn into a conscious thought and then drive a behavior." So looking at this process, it may seem as though we are helplessly chained to the elephant of our emotions. A part of this is true, but the rider can influence the elephant's automatic emotional responses in significant ways over time. Change the way you perceive the world around you, along with changing the automatic feelings and thoughts coming up within you, and you'll find that what you choose to attend repeatedly and over time will harness and refine the internal elephant.

The default mode of the human brain is more sensitive to negativity. This is part of the built-in survival system of the human race. Being hyperalert to negativity in the environment—say, a noise that sounds like a bear in the forest—was one way to assure survival in earlier, more dangerous times. Now that humans live in a world that is for the most part safer, the brain still has a residual focus on negativity, but it no longer serves us well.

Here's an example. You have a headache. Let the brain take its natural negative course, and it begins to spiral into worst-case scenarios faster than you can say, "Gee, I must have a tumor." Within a few seconds your inner elephant can take you down a path that imagines your headache is a baseball-sized cancerous mass, skips straight to visualizing yourself on your deathbed, fast-forwards to imagining your own funeral in living color (complete with songs, hymns, flowers), and gets angry at your spouse when you contemplate him or her marrying someone else you don't like. At this point your mate may innocently walk in the room, and you find yourself feeling and acting inexplicably ticked off. This downward spiral can take place within seconds. This is how quickly a response to a trigger (a headache) can deteriorate into a negative spiral, instigating bad behavioral outcomes (you are unfairly angry and short with your spouse for an imaginary future scenario).

The good news is that we can also, and just as easily, spiral into positivity.

TAKING THE EMOTIONAL REINS

Everything that comes into our brain, be it from external sound or visual cues or memories that pop into our mind, causes an automatic emotional response. Our brains begin processing things like body language or the tone of someone's voice, and this leads to emotions that are subconscious reactions. But as these emotions become more conscious they turn into feelings. Butterflies in your stomach would be one example. At this point we can consciously switch what we are attending to and choose to focus on better and more positive thoughts.

An example might be that you are about to speak to a group in public. You realize that your mouth is getting dry and your stomach is flip-flopping. The automatic fear response is in full swing. However, this is when you can choose to "attend to" more positive thoughts and actions. You can begin to breathe slowly and deeply. The brain takes a cue from this and begins to relax. You can think about how much your audience needs and wants the information you have to share, taking the focus off your fear and onto their receptivity instead. Continue like this and before long you will be relaxed and positive, looking forward to stepping up to the podium with energy, focus, and joy.

The rider has tamed the elephant.

Another scenario: Your inner elephant really wants a sugary cookie and would like to go down the path leading to an extreme sugar high. But you've also trained your brain to pause and recall the emotions that come after the sugar cookie. The fatigue, the shakes, the excess weight. You remember that after the high comes the crash. You've begun to train your brain to want more positive outcomes, so you choose instead to eat half of a frozen banana that has been rolled in chopped nuts, stored ahead in your freezer for such a time as this, going down the path to a truly satisfying and healthy snack. This healthy treat satisfies your urge for something sweet while also giving you fiber and protein and potassium and more. You know you will feel better and won't have a headache or be hungry again in thirty minutes. You give your body what it *really* wants and needs, satisfying both elephant and rider.

Now imagine that you practice positivity in every circumstance where you notice your thoughts about to spiral down. You stop them and restart a more positive spiral and change your mood state, and then automatically better behaviors will follow as well. You do this so often that it becomes habitual. Over time you can even change your personality. You can go from a fearful, worst-case-scenario auto-responder to a positive, happier, more relaxed, productive, and enjoyable person. You make better decisions, both in the short term and long term. No matter your age, you can do this.

Besides being aware of what to pay attention to and focus on, there is another way to speed up and help your brain integrate and respond to life in better ways. It only takes ten to fifteen minutes, three times a week, sitting in front of your computer and playing a few fun and relaxing "games" in our 24/7 Brain Gym on the Amen Solution website (www.theamensolution.com).

MENTAL WORKOUTS

Think of training the brain to respond in more focused, positive, and calm ways as we think of exercising for our bodies. I like to call it mental workouts. And just as we exercise, eat right, and brush our teeth as a form of preventive health, by training the brain on a regular basis we are practicing preventive mental health. Should a crisis come along, you'll have trained your brain to deal with stressors and problems more effectively. Better responses to all of life's challenges will have become habitual, so you are not as easily thrown into an unproductive negative loop.

We are all constantly on a thin edge, which can lead us down one side to negative spiraling or another side that can lead us to more positive and helpful reactions and behaviors. The brain is also extremely suggestible and open to cues and clues. For example, if you are feeling insecure, try standing up tall with shoulders straight and head up, and smile with confidence. Adopting a confident body posture sends a message to your brain that says, "I'm feeling confident that I can tackle this challenge." Latching onto role models of confidence, listening to audiotapes, being around confident people, and reading quotes that support a more confident state of mind can all affect which side of the "edge" your brain decides to ski down, which in turn affects your behaviors in the real world.

The 24/7 Brain Gym provides exercises that keep your brain tuned up and your automatic habitual responses to life's many triggers more positive, hopeful, and calm. It covers these four areas:

- Boosting your memory and attention
- Enhancing your emotional IQ
- Increasing your happiness
- Reducing your stress

In addition, the site offers:

- Emotional training games to help you to better pick up on non-verbal clues
- Thinking training games to help boost your attention, memory, and planning skills
- Feeling training games to help minimize stress and enhance your heath and well-being
- Self-regulation training games to help you manage your emotions, thinking, and feeling

Together these four areas help keep your brain in top shape, so both the elephant and rider parts work in tandem and allow you to handle life better. This is especially important when it comes to longevity. Those who visualize themselves as getting better as they grow older, who look forward to their golden years, and who respond to life's adversities with resiliency and positivity indeed live longer and happier lives. One of the testimonials from a client who used the Brain Gym training was that when her father passed away, she felt she was better able to cope with the grief and changes than she would have been before she'd toned up her brain. She felt it helped her to be more effective with difficult decisions and comforting others who were also grieving, and she was more resilient post-loss.

The games in the Amen Solution Brain Gym are amazing—and fun. Even in the first week I could see my abilities improving. What an innovative way to increase intelligence, emotional IQ, and internal self regulation!

—Bill Harris, creator of Holosync

Consider adding mental workouts to the rest of your longevity habits, and it will pay you back many times with increased feelings of calm, happiness, and focus. You'll also be proud of the life legacy you leave others as your outward behaviors more and more line up with your inner convictions.

DO-IT-YOURSELF BRAIN IMPROVEMENT

No matter your age, income, IQ, or education, there are dozens of ways to help your neurons grow, stretch, and branch into a younger, more beautiful brain every single day. Here are a few examples:

1. Learn a new language. Learning a new language requires that you analyze new sounds, which improves not only auditory-processing skills but also memory.

2. Play Sudoku. Sudoku is a numbers (not math) game that is both popular and addictively fun to many who play it. It can help increase your logic and reasoning skills as well as your memory. Crossword puzzles do the same.

3. Lose the list. Using mnemonics (triggers to aid memory using visual imagery or sounds, such as rhyming) is a great way to boost

your brain while developing a system to remember things. There are several great memory courses available on audio or video recordings, often at local libraries or online.

4. Get in the game. Play board games like chess or Scrabble. Trivia games can boost memory, jigsaw puzzles can help visual and spatial skills, and mah-jongg can help executive function (the capacity to control and apply your mental skills).

5. Online brain-training games such as our Brain Gym at www.theamensolution.com can be quite helpful in keeping your brain fit. Spend about ten minutes a day doing these fun games, and see if you don't find your brain beginning to process better and faster.

6. Be a Curious George. Stay curious about life and learning. Read and study or take courses in subjects or the arts or activities that capture your fancy. Be a lifelong learner and you're more likely to stay young at heart and in your brain.

7. It is never too late to go back to college! "People with fewer academic qualifications may grow old faster," according to a DNA study that compared groups of people who spent different lengths of time in education and found the ones who spent the least time had shorter telomeres, or "caps," on the ends of their DNA, a sign of premature aging in cells. Think you're "too old" to earn a degree? Ask yourself, "How old will I be in four years if I don't earn a degree?" The oldest person to graduate college in the United States was in her midnineties! Already have a degree? How about getting another one? Or go for a variety of continuing education courses, designing your own degree in "What I've Always Wanted to Learn."

8. Learn to play a musical instrument or a different instrument than you normally play.

9. Try a brain healthy sport you've never tried.

10. Try a new brain healthy recipe, perhaps from one of my wife's cookbooks.

11. Break your routine. This is especially important for anyone who is tethered to bad, brain-harming habits. You can increase your chances of staying healthier longer if you change your daily habits and routines. Introducing new habits can help rewire your brain so you don't fall back into the same patterns of activity. For example, if you always take the same route home to work so you can stop at your favorite doughnut shop along the way, take a different route to work and bring a homemade brain healthy smoothie made from protein powder and fruit, which you can sip along the way.

SPECIFIC WORKOUTS FOR DIFFERENT BRAIN AREAS

Here are some workouts I recommend to help balance six different areas of your brain.

- PFC (forethought)
  - o Strategy games, such as chess and checkers
  - o Meditation to boost PFC function
  - o Hypnosis, which can help focus and boost PFC function
- Temporal lobes (language and memory)
  - o Crossword puzzles and word games
  - o Memory games

- Basal ganglia (modulate anxiety and motivation)
  - o Deep relaxation and/or meditation
  - o Hand-warming techniques. As you warm your hands it sends an automatic signal to the rest of your body to relax.
  - o Diaphragmatic breathing
- Deep limbic (emotions)
  - o Killing the automatic negative thoughts. There is more information in chapter 7.
  - o Gratitude practice
  - o Building libraries of positive experiences to enhance mood states
- Parietal lobes (direction sense and spatial orientation)
  - o Juggling
  - o Interior design
- Cerebellum (coordination)
  - o Dancing
  - o Table tennis (also works prefrontal cortex)
  - o Martial arts, without risk for brain injury (also works PFC and temporal lobes)
  - o Handwriting
  - o Calligraphy

How old would you be if you didn't know how old you are? By keeping your brain young, curious, and ever learning new things in this fascinating world of ours, you may find yourself growing younger, rather than older, as the years go by.

CHANGE YOUR AGE NOW: TWENTY BRAIN TIPS TO HELP YOU BECOME A LIFELONG LEARNER

1. To find your motivation to learn something new, begin by asking yourself, "What gifts do I have to bring to the world that could be

lying dormant between my ears?" Then ask, "What do I dream of doing with my life?" Go ahead, start writing your bucket list now. Your brain loves activities that hold promise and excitement like this one.

2. Think of books as a college course between two covers. Books are the world's greatest educational bargains. You can learn from the world's greatest minds, past and present (and for little cost), if you become a reader. You can become an expert at almost anything, at any age!

3. Turn your car into a "university on wheels." Purchase, download, or borrow audiobooks on a variety of subjects that pique your interest. Download podcasts from great teachers whom you admire. You'll turn boring drive times into classrooms of fascinating knowledge.

4. Stop telling yourself you have a poor memory or that you are not a good reader. Instead say, "Memory is an art I can practice. I can read as well as anyone by applying new habits."

5. "Don't let your schooling get in the way of your education." Formal learning is important for many, but only those who go above and beyond the educational system discover the true joy of lifelong learning.

6. When memorizing a list, associate it with the craziest picture you can think of to help your brain recall it later. No one sees the image you are holding in the privacy of your mind, so be creative and have fun with it.

7. To remember someone's name, repeat the name, use it once or twice in natural conversation, visualize the name as a picture

(perhaps on the person's forehead), and use their name when saying good-bye.

8. Try increasing your reading speed by using the simple method of following sentences with your finger, a pencil, or your mouse cursor.

9. Set aside three or four ten-minute sessions a week to play a variety of brain games on your computer. It's like circuit training for your mind. We offer many great exercises for your brain in the form of fun games in our 24/7 Brain Gym at www.theamensolution.com.

10. Sharpen your brain by enjoying leisure activities that also keep you thinking. A study published in the *New England Journal of Medicine* found that reading, playing board games, playing musical instruments, and dancing were among the best leisure activities for keeping your brain young.

11. Try breaking your routine by doing something different and outside the box. Try a new sport. Whip up a new recipe. Take a new route home. Mix up your life. Variety is not only the spice of life, but it will help grow new neurons in your brain too!

12. When you really want to learn something well, make sure to teach it to someone else. This will dramatically increase your skill and knowledge in a subject within a short period of time.

13. Note-taking increases comprehension and retention. When you read, take notes in the margin of a book. If you're reading a book that you cannot mark up, keep a notepad handy as you read. Jot notes on the strips of paper, then tuck them in the pages where you found the quote or point you liked.

14. Lifelong learning involves becoming more curious about all of life. For example, to keep your mind from wandering in conversations, be a curious listener. Notice not just what another person is saying but also her body language and tone. Ask follow-up questions. Pretend you are a journalist or a therapist, deeply intrigued with the story you're being told and the stories behind the story.

15. Don't ask yourself, "How smart am I?" Instead, ask yourself, "How am I smart?" There are many kinds of intelligences: social, mathematics, logic, art, creativity, intuitive sensing. What do you excel at?

16. Everyone has preferred learning styles. Find your style. Do you learn best by reading, hearing, talking, writing, doing, or some combination of these? Try to learn something new via your best learning style. If you are an auditory learner, listen to a book on tape. If you are a kinesthetic learner, take a class where you'll have hands-on experiences.

17. Cognitive skills tend to dip after we graduate from college or retire from work. Don't stop challenging your brain on a daily basis! Be a perpetual student of life. Take continuing education classes or get a college degree. Learn to be a gourmet cook, discover fly fishing, write your memoir, study the brain! The world is endlessly fascinating for those who never stop learning, and it helps your brain thrive.

18. Meditation has been shown to boost activity in the PFC and sharpen your mind. Just a few meditative minutes a day can make a dramatic difference in your mental abilities.

19. Boost your prefrontal cortex by making clear goals and looking at them every day.

20. Boost your brain's flexibility and creativity centers by asking yourself to look at everyday activities, such as family time or how you do an activity at work, in new and different ways.

6

JONI AND THE MINI FACE-LIFT

BOOST YOUR BLOOD FLOW FOR BETTER SKIN AND SEX

Good morning, Daniel . . .

Just a quick note to tell you that my husband & I, along with our granddaughter, were walking around Laguna Lake where I stopped to read a sign that read "Please Don't Feed the Ducks." It went on to warn people not to give them a handout because they need to feed off of their natural food plants and insects. If the ducks feed off of the food we hand out it could be deadly, as it changes their behavior. They become sedentary and will stay at the lake instead of migrating.

It was a huge wow *moment for me. When I don't eat the foods created for me and I eat food handed out (the drive-thru window!)—"fast food"—it changes my behavior. I will not "migrate" to new horizons!*

I just had to share this with you because it really is a parable about all that you and your wife are so passionate about. Oh, by the way, the sign said when we feed the ducks, other ducks come to the lake and it causes unsightly droppings. So the moral of the story is when junk food is served you sit around, with no motivation, in your own &%@!

Have a grand day!

Joni

JONI NOW: BEAUTIFUL INSIDE AND OUT

Joni is a woman whom I admire for her enthusiasm, unique way of seeing life, ready sense of humor, and ability to encourage others. Over the years she has sent me a number of e-mails like the one on the previous page. Recently Joni started attending classes at the clinic in Newport Beach about how to have a better brain and a better body.

"I started applying what I learned. Eating better, taking supplements, and exercising more consistently. I think consistency is a challenge for women when we're raising kids because we tell ourselves, 'I don't have time to exercise! There are too many needs I have to tend to!' But what we don't realize is that taking time to exercise is not only an investment in ourselves but also in our family members, who want us to live longer and feel good about ourselves."

Indeed, a happy, healthy wife and mom is one of the best gifts a woman can give her husband and kids. Recently Joni, who is a grandmother of three, shared the following story. "One day after I'd lost about ten pounds and started to feel much better in every way, I went out to breakfast with my twin sister. She eyed me suspiciously across the table and said,

"'Okay, tell me the truth, Joni. Did you have a mini face-lift?'

"I was so surprised, and flattered. 'No!' I said, 'I promise I haven't!'

"'Seriously?' my twin countered. 'Even my sons thought you'd had some work done!'"

Joni told her sister that all she had done was to change her diet, add supplements, and exercise consistently. "It was only a ten-pound weight loss, but it was more that the pounds lost gave my skin a lift and a happy glow."

In truth, the process began fifteen years ago when Joni was, as she simply put it, "a mess." She explained, "Transformation for me personally came from the inside out. Outside in has never worked for me.

"I am sure that in addition to the dietary and exercise changes, the reason my skin and face appear younger is that I am happier and more peaceful and balanced within."

JONI THEN: A HOT MESS

In 1996, when I first met Joni, she thought, as many people do, that her issues stemmed from a lack of willpower. "I'd pray and pray and ask God to help me, but I wasn't getting much better. Until, that is, I also started therapy and that wise therapist sent me to Dr. Amen."

Joni had a SPECT scan at our clinic and it was a huge epiphany for her, as it is to so many thousands of patients who come face-to-face with the first picture of their brain. At the time she was terribly anxious and depressed. She had been diagnosed with PMS, but she knew it had to be more than that. She was living in such pain, misery, and shame. She had been emotionally wounded as a child, as so many women are. She believed she was dumb and had such a hard time feeling that she belonged anywhere in the world. As she was sitting in our waiting room, she was engulfed with a feeling of shame for being such a mess that it had come to the point of literally having to get her head examined.

DISCOVERING HER INNER BEAUTY

When she saw a picture of what was happening in her brain it was as if she just opened up inside.

Joni said, "As I looked carefully at the images I really understood that something biological was at play here. This wasn't just a spiritual problem or lack of willpower. The doctor, pointing to various parts of my brain, described why I was feeling so anxious. My basal ganglia were overactive. When I saw low activity in my prefrontal cortex, tears of compassion welled up in my eyes. Compassion for my lost self. I could see that without enough blood flow to this part of my brain, it was no wonder I struggled in school, with my marriage, and even feeling loved by and connected to God."

GETTING BEAUTIFUL ALL OVER

The only plastic surgery I have had is cutting up my credit cards!

—Joni

In addition to the therapy, having her brain balanced with supplements and a very little bit of medication—a small amount of Adderall for the ADD issues—gave her a brain she could finally work with. Therapy continued but much more effectively. Once she was better on the inside, she was fully ready to get her body healthy as well. About this time my book *Change Your Brain, Change Your Body* came out. The clinic offered classes and a support group and she signed up and brought my daughter's friend, who lost thirty pounds.

The classes helped her understand the connection between what she put in her mouth and her mood. Healthy food could bring a full life, instead of an overstuffed belly. In addition to upsizing her nutritional input, Joni found that targeted supplements were very helpful.

"I'm going through menopause and that is no picnic, but I'm so grateful to have the supplements, the nutrition, and the exercise to help me," she said. New research suggests that fish oil lowers the frequency of hot flashes. Joni works out with a trainer three times a week. She shared that before she started on her supplements she used to come in to the gym to work out and forget what her trainer had told her to do. Her trainer would tease her, saying, "What are you doing, Joni? Waiting for a bus?"

To stay on task was that difficult for her. But with the supplements and nutrition she has the focus and energy to pay attention and finish her workouts. She also goes on hikes and bike rides.

When she goes to the grocery store, Joni is now an avid label reader.

She realizes that food is truly medicine. She's also getting comfortable with accurate portion sizes. Because Joni has ADD, she sees quite an improvement in her attention when she has protein for breakfast. She usually goes for a breakfast shake with protein or egg whites with avocado and pico de gallo. For lunch she usually has a soup and salad or a healthy sandwich. Dinner is typically fish, a vegetable, and a salad.

Her time of greatest temptation is at night, when she stays up late. Her brain profile showed Joni struggles with impulsive and anxious overeating. "Every day I have a choice," she says. "Interestingly it wasn't just food I was impulsive about. I used to be an impulsive shopper as well and got in trouble with going to the mall and ended up with out-of-control credit card bills. Now I'm debt free! As I got my ADD balanced, I had better control over eating, and many other areas of my life."

Joni added, "In response to the mini face-lift and debt reduction, I tell people, 'Honestly, the only plastic surgery I have had is cutting up my credit cards!'"

Joni says that her supportive community made all the difference in being able to stick with her brain healthy decisions. "I have people in my life who support my healing and also people I can call when I'm triggered or stressed. My brain scan also showed a pattern consistent with past emotional trauma. Now I understand what happens when something triggers old pain, and my brain wants to do something to self-medicate, like shopping too much, eating a bunch of sweets, or other addictive behaviors. I know how to soothe my brain now, with supplements and diet and exercise and changes in thought patterns.

"The other day someone came into our workplace and just exploded with anger. This was a trigger to major anxiety for me, and I really wanted to 'medicate' it with a bunch of doughnuts. Food, to me, was like a loving mama, like Paula Deen from the Food Network!"

What Joni has realized now is that she can get that comforting and support and nurturing from Paula Deen–like people in her life—without all the sugar and butter to go with it!

"I'm thrilled with the reflection I now see in the mirror," says Joni. "What is within me now shines on the outside! With a balanced brain I no longer see myself as I used to see me, like the wicked queen in a fairy tale gone wrong. What I see reflected in the mirror now is a healthy, vibrant young woman!"

CARING ABOUT YOUR SKIN IS IMPORTANT

"Beauty is in the eye of the beholder" is an old saying that may be more true than we realize. The "beholder"—the human eye—automatically sizes up someone's health by the appearance of their skin and face. Surprisingly, this automatic response is amazingly accurate. Early visible signs of aging, such as wrinkling, often indicate some sort of systemic failure within. A growing body of research demonstrates that facial wrinkles are a reliable indicator of internal health. For example, in one study using pictures of older twins, the researchers had twenty nurses browse the photographs to guess the age of each individual twin. In the following years, the twin who was seen as "older looking" was more likely to die from health-related issues in 73 percent of the cases.

One of the most important concepts that I have learned over the years is that *the health of your skin is an outside reflection of the health of your brain.* The brain–skin connection is so strong that some people have begun calling the skin the brain on the outside. The same habits that we have been discussing in this book that improve the brain's look, feel, and function, also apply to your skin. Your skin covers about twenty square feet and is one-sixth of your body weight, making it the largest organ in the body. Nourish it well, starting from the inside out, and it will pay huge dividends in both the quantity and quality of your life.

Did you know that 50 percent of the brain is dedicated to vision? Healthy skin is attractive, not only to the opposite sex, but to everyone. Many people think it is shallow to care about how you look, but it is important to take care of your appearance if you want to stay connected to others and if you want to attract people to you.

"BUT WAIT! DON'T GET THAT FACE-LIFT YET!"

Extreme makeovers are fun to watch on TV, but this kind of beauty is only skin deep at best, doing nothing to help a person's health and longevity. The long-term results are fleeting.

What you don't see with some of these extreme makeovers is the extreme *pain* some of the procedures involve. In addition, few nip 'n' tucks are without side effects. Many times the results are less than people had hoped for, or worse, they are left with a face stuck in the "surprise" or "Joker-face smile" position. Google "bad celebrity face-lifts" and you'll get a brief tour into the bizarre. Surgery of any sort, including implanting or injecting foreign substances into the body, always carries both short-term and long-term risks.

Sure, there are cases when plastic surgery is a legitimate option. However, if you've been toying with the idea of going under the knife for some "work," I'd like to challenge you to consider an alternative. Postpone the decision for six months to a year, and in that time make a concerted effort to make positive changes that will uplift your health, your happiness, and your face at the same time. As another added bonus, doing this will not only tighten and smooth the skin on your face but on the rest of your body as well.

At the Amen Clinics we often see people morphing into younger versions of themselves as they get their brains balanced by going on a brain-smart diet and exercise program. Even a ten-pound weight loss—if you are losing weight on highly nutritious foods—can often de-age a person's face profoundly. When people lose weight on a healthy regimen, others often comment on how much younger they look as well (as opposed to those who lose weight by cutting calories but eating low-nutrient, artificial foods—these people tend to look drawn, wrinkled, and weak).

I remember when I once tried an extreme low-carbohydrate diet: my skin took on a grayish tinge, and I looked older and sicker in a matter of weeks. It was the wrong weight loss for me.

There's a good chance that if you choose brain healthy living, and dive into it wholeheartedly, you will look in the mirror a few months from now and see significant progress. But be sure to take a before picture when you start. Sometimes gradual changes are more noticeable to others.

In the rest of this chapter we'll explore natural ways to lift, tighten, and smooth your skin. We will also discuss how to boost the blood flow to your brain and skin and improve your sex life.

As it turns out—surprise, surprise!—what keeps your brain young also keeps your skin young. So you get a two-fer when you opt for a brain healthy life. Actually, you get a four-fer, because what is good for your brain is good for your skin is also good for your heart and your genitals. There is no downside, no side effects, no risks, and no frozen clown faces when you choose to go for a natural, brain healthy face-lift!

> There are no beauty contests for a beautiful brain, but the truth is that without a gorgeous brain, you will not be as good-looking as you could be. Anger, depression, and anxiety all show up on our faces. Getting the brain calmed and focused will relax and beautify your face as well.

HEALTHY SKIN FROM THE INSIDE OUT

Even the best facelift does little if you are not addressing the aging that is going on inside the body. To truly look and feel young, you have to start from within and, most important, with the brain.

—Dr. Eric Braverman

One of the things I love so much about Joni's story is that she emphasizes that her changes on the outside began with changes within. This applies to many areas of life, but it is perhaps even more obviously true when it comes to our skin. Caring for our skin is a little like caring for a plant. A well-nourished plant grows tall and beautiful, with supple leaves and bright color. We know this about plants. But we forget this about ourselves.

Imagine this scenario.

Let's say a woman named Sophia has a beautiful flowering plant that she absolutely loves. But one day she thinks to herself, "Gee, I'm tired of watering and fertilizing that thing. And I prefer to move it to a dark corner in the basement. Also, I think I'll try growing it in clay dirt instead of potting soil. And forget the Miracle-Gro. This year I'm going to save myself time and money by letting this plant take care of itself."

So Sophia quits watering the plant, lets the soil go without nutrients, and keeps it away from sunlight and fresh air. Her plant is basically relegated to solitary confinement, void of nourishment and light, without even bread—or rather, fertilizer—and water.

Eventually the inevitable happens. The leaves begin to shrivel, wrinkle, and turn brown. Sophia doesn't like this look, so she trims the dead leaves and spray paints the plant a nice shade of "grass green."

The stems begin to bend, so she props them up with sticks and tape and wire. Even so, the root and stems of the plant's vascular system have bends and kinks aplenty that keep nourishment from flowing freely to the flowers and leaves.

The plant no longer blooms, so she buys artificial flowers and "implants" them among the leaves.

"Wait! Wait!" you would say to anyone trying to do this sort of extreme makeover to a once naturally beautiful plant. "You're killing that poor plant! It has to be nourished with rich soil, daily watering, and just the right amount of fresh air and sunlight. Then its stems will straighten up, the leaves will be supple and vibrant, and real, colorful flowers will bloom."

The parallel is obvious. You can't grow a healthy, beautiful flower by such extreme and artificial means; and neither can you grow a healthy, supple, unwrinkled, vibrant skin with color blossoming in your cheeks unless you "grow it" from the inside out.

The most expensive cosmetics in the world cannot make you as beautiful as good brain-smart living can. Let's take a look at how you can begin to nourish your skin, starting now.

WATERING YOUR SKIN: HYDRATION

Just as plants need water, the skin and brain also need water. Drinking plenty of filtered water is most helpful for the skin because of its ability to flush toxins from the body, including those that end up in your skin. Green or white tea might be even more helpful, as it is shown to protect collagen.

HEALTHY OILS

Omega-3s A study in *Journal of Dermatological Research* (2008) showed an improvement in skin elasticity in women taking over 1 g EPA (eicosapentaenoic acid) fish oil for three months. Omega-3s also help the skin recover more quickly after exposure to UV radiation and protect DNA within skin cells along with supporting collagen. It also can improve the elasticity of blood vessels, thus improving blood flow, giving your skin a healthy glow.

Gamma-linolenic Acid (GLA) GLA acid is a "good" omega-6. It is typically derived from borage, evening primrose, or black-currant seed oils. (Borage contains the highest amount.) Improvements in skin smoothness and hydration have been shown after one to three months of consuming 500 mg GLA from borage oil; it also helps relieve dry, itchy skin. By combining with fish oil, you'll reduce the amount of water lost from the epidermis. Give this combination a try for at least a month, as it takes this long before effects begin showing up on your skin.

Acetylcholine This is a very important nutrient that acts as a neurotransmitter in the brain, for learning and memory, but also in the skin. Acetylcholine provides the moisture to the body, so when it goes, we dry up from the inside (starting with the brain). Dry, wrinkled skin may be a clue that your memory is fading as well. A loss of acetylcholine is a marker of dementia.

Here are some foods that help increase acetylcholine:

- Whole egg
- Turkey liver
- Cod, salmon, or tilapia
- Shrimp
- Soy protein
- Peanut butter
- Oat bran
- Pine nuts (Be sure these are grown in America, as pine mouth syndrome, a temporary condition affecting your taste buds, has been found in many imported batches of pine nuts in recent years.)
- Almonds
- Hazelnuts
- Macadamia nuts
- Broccoli
- Brussels sprouts
- Cucumber, zucchini, lettuce
- Skim milk
- Low-fat cheese
- Low-fat yogurt

Supplements are also a great way to increase your acetylcholine. These include phosphatidylcholine, acetyl-L-carnitine, alpha-lipoic acid, manganese, and huperzine A.

Tea Time For healthy skin that glows from the inside out, drink tea: white, green, or black. Studies show the polyphenols found in tea have anti-inflammatory properties that may be beneficial to skin. One study published in the *Archives of Dermatology* found drinking 3 cups of oolong tea a day cut eczema symptoms for 54 percent of those who tried it. Research found that drinking two to six cups of green tea a day not only helps prevent skin cancer but might also reverse the effects of sun damage by neutralizing the changes that appear in sun-exposed skin.

There is also some compelling new research that shows white tea to be a strong skin de-ager, in addition to helping reduce the risk of cancer and rheumatoid arthritis. In a study where scientists tested the health properties of twenty-one plant and herb extracts, they were intrigued to find that white tea considerably outperformed all of them. White tea had antiaging potential and high levels of antioxidants, and protected the structural proteins of the skin, specifically elastin and collagen, which help the skin's strength and elasticity, diminishing wrinkles and sagging. As little as one cup of white tea a day has a powerful effect.

FEEDING YOUR FACE

You may have noticed that Japanese women have some of the most beautiful skin in the world. Often they seem almost ageless. Many of the world's longest-lived people are from Japan, and their skin shows the least signs of visible aging. Researchers believe it may have much to do with their diet, which is high in seafood and vegetables while low in sugars and bad oils. Just as a plant looks best when it is "fed" rich soil and nutrients, so a face that is "fed" healthy food will look its best too.

In one study, researchers found that the older adults with the highest intake of items like olive oil, fish and seafood, nuts, legumes, yogurt, tea, whole grains, deep green veggies, and dark fruits and berries had the fewest visible signs of aging. On the other hand, a greater risk of skin wrinkling was associated with the consumption of fatty processed meats, saturated fats, white potatoes, and sugary beverages and desserts. Your skin is what you eat, so feed it beautiful ingredients.

Dr. Lawrence E. Gibson, dermatologist for the Mayo Clinic, lists the following foods as best bets for healthy skin.

- Carrots, apricots, and other yellow and orange fruits and vegetables
- Blueberries
- Spinach and other green leafy vegetables
- Tomatoes
- Beans, peas, and lentils
- Fish, especially salmon
- Nuts

Making sure you get plenty of fiber-rich foods also slows down sugar spikes, which reduces glycation, a process that ages the body, including the skin.

SUPPLEMENTS

More and more multivitamin products are being formulated with skin-nourishing ingredients and advertised as such. Look for the following ingredients in your skin-friendly multivitamin:

Vitamin C and Lysine These antioxidants are known to inhibit enzymes produced by cells that are known to attack collagen.

Vitamin D This important vitamin supports skin health by decreasing inflammation and boosting immunity. As skin ages it is less effective at synthesizing vitamin D coming from the sun. If your skin is not healing as quickly as it used to from small cuts, extra oral vitamin D may help.

Zinc This mineral promotes wound healing and is beneficial to hair and skin health, especially if you suffer from dermatitis and dandruff.

Vitamins A and E Both of these vitamins have been beneficial in decreasing inflammation.

Omega-3s Omega-3s improve skin elasticity, help protect and restore body after UV radiation exposure, and protect DNA and collagen. As previously mentioned, GLA can help as well and, taken in combination with an omega-3 supplement, can, over time, greatly improve the moisture and appearance of the skin.

Dimethylaminoethanol (DMAE) Also known as deanol, DMAE is an analog of the B vitamin choline. DMAE is a precursor of the neurotransmitter acetylcholine, and it has strong effects on the central nervous system. DMAE is commonly used to increase the capacity of neurons in the brain and is also thought to have antiaging properties that diminish wrinkles and improve the appearance of the skin.

Phenylalanine This amino acid has been found to be helpful for depression and pain. There is also good scientific evidence that it may be helpful for vitiligo, a chronic, relatively common skin disorder that causes depigmentation in patches of skin. It occurs when the cells responsible for skin pigmentation die or become unable to function.

Alpha-lipoic Acid This compound is made naturally in the body and may protect against cell damage in a variety of conditions. In a number of studies it has also been found to be helpful for skin issues as well.

Grape Seed Extract This substance comes from grape seeds, which are waste products of the wine and grape juice industries. Extensive research suggests that grape seed extract is beneficial in many areas of health because of its antioxidant effect to bond with collagen, promoting youthful skin, elasticity, and flexibility.

Probiotics The gut is often an indicator of health, and this is especially true when it comes to stress. Did you know that the emotions of fear and anger can reduce the number of healthy bacteria in your gut tenfold? In times of sickness or emotional stress, or after overindulging in alcohol, you help your body get better more quickly by ingesting fermented milk products like Kefir or yogurt or taking a good-quality probiotic formula. Probiotics can help in returning to normal immune functioning after exposure to UV rays. Animal studies (yet to be conducted on humans) show promise for skin rejuvenation following exposure to sun by use of probiotics.

> Splenda (sucralose) was found to reduce friendly bacteria and also to change healthy pH balance in the body, so avoiding it as a sugar substitute is a good idea. I recommend stevia instead.

HORMONE BALANCING

Having your hormones checked and balanced is critical for both brain and skin health. Trouble with testosterone, estrogen, progesterone, thyroid, and cortisol levels can cause major problems. But before you start taking synthetic hormone replacement, clean up your diet and environment first. For example, did you know that getting a sugar burst can lower your testosterone levels by 25 percent? Testosterone is thought of as the libido hormone. This means if you share the cheesecake at the restaurant, no one will get dessert when you get home. Estrogen replacement for women, when appropriate, can slow the aging process in the skin by slowing collagen loss that affects the skin's ability to retain moisture.

SUNLIGHT: JUST ENOUGH, NOT TOO MUCH

Most people need about twenty minutes of sun a day to get a healthy level of vitamin D. You can also get the necessary amount with a vitamin

D_3 supplement if you live in a part of the country, like Washington or Oregon, where there are months where the sun hides its face. Avoid the harshest hours of sunlight, however, between 11:00 a.m. and 2:00 p.m. A variety of plants need differing amounts of sun and shade; this is also true of people. Those with light blond and red hair and fair complexions need to be especially careful of sun exposure, as their rates of skin cancer are much higher than the rest of the population with olive or darker skin tones.

To Sunscreen or Not to Sunscreen? Years ago, after researchers discovered what overexposure to sun could do to our skin, the general advice was to slather yourself up with a strong SPF sunscreen. What they didn't take into account was that some of the ingredients in some sunscreens may actually be more harmful to the body than reasonable sun exposure. You can find a list of the safest sunscreens to use at the website of the not-for-profit Environmental Working Group (www.ewg.org). They also did not take into account that as a nation our vitamin D levels would drop to dangerously low levels. The answer here, I believe, is to be cautious with the sun. Get enough to keep your vitamin D levels healthy but not too much to put your skin health at risk. Think twenty minutes a day without sunscreen. Never get sunburned, and get your vitamin D levels checked with a test called 25-hydroxy vitamin D level.

Get a Healthy Tan from Your Five a Day Scientists have shown that you can give your skin a golden glow by eating more fruits and vegetables with carotenoids, which will subtly alter your skin color. Carotenoid, an antioxidant responsible for the red coloring found in, for instance, tomatoes, peppers, plums, and carrots, eventually imbues the human skin (of Caucasians) with a healthy-looking golden glow. Carotenoids are stored in fat under the skin and secreted through the skin in serum. Then they are reabsorbed into the top layer of the skin, bestowing that golden color. These substances also contain powerful antioxidants that are good for your brain's health. Studies show that people find a healthy golden glow more attractive than lighter pigment. Researchers believe that rosy skin tones slightly flushed with blood and full of oxygen make people look healthier, because they suggest a strong heart and lungs.

GOOD VASCULAR HEALTH EQUALS HEALTHY SKIN

It's all about blood flow. The skin contains an abundance of blood vessels, making it vascular in nature. These blood vessels promote circulation, blood flow, and clean the skin. When our cardiovascular system clogs up and blood flow is poor, skin loses its pink, youthful appearance. Besides wrinkling, older people often have sallow, pale skin because of a lack of blood flow.

Take deep breaths of fresh air. Put a plant under a glass dome and it won't last long. Living things need oxygen and lots of it! Deep-breathing exercises help calm the body and increase oxygen to the blood. Breathing outside, in nature, particularly where there are lots of trees (what the Japanese term "forest air breathing") lowers cortisol, decreases glycation (a part of aging), and improves well-being.

Avoid anemia. Anemia can cause your healthy red blood cell count to drop too low and cause people to look ghostlike. Red blood cells contain hemoglobin, which carries oxygen to your tissues, and a lack of oxygen can cause stress on organs. Signs of anemia can be mild skin paleness and dizziness or lightheadedness.

Increase blood flow to your skin. The best way to get the blood moving is through exercise.

1. **Exercise Makes You Sweat** "The body only has so many mechanisms to rid itself of toxins—the kidneys, the liver, and the skin," says Sandra M. Johnson, M.D., a dermatologist with Johnson Dermatology in Fort Smith, Arkansas. "Exercise increases blood

flow to the skin, increases neuronal stimulation, and allows the sweat glands to increase their functions and rid the toxins." Once you sweat out those toxins, be sure to wash them off. Showering after a workout keeps them from sitting on your skin, preventing bacterial or fungal infections that may occur from dirt clogging up your pores.

2. **Exercise Tones Your Muscles** The more toned you are beneath your skin, the healthier your skin will look and feel. Toning your muscles may also help to minimize the appearance of cellulite. You can't exercise cellulite away, but you can help it look better.

3. **Exercise Boosts Oxygen and Blood Flow to the Skin** Studies have shown that regular exercise boosts blood flow enough in type 2 diabetics to help reduce the risk of skin problems that lead to amputation. Exercise increases blood flow, which means that more oxygen and nutrients are being carried to the skin's surface.

4. **Exercise Eases Stress** Exercise has long been known as a great way to relieve stress. Those mind–body benefits may extend to your complexion as you frown less, smile more. Also, some skin outbreaks may be stress-related.

5. **Exercise Gives Your Complexion a Beautiful, Natural Glow** When you exercise, your skin begins to produce more of its natural oils that help skin look supple and healthy.

    Exercise is a vital secret skin care ingredient for a young, healthier, and smoother complexion.

Treat sleep apnea. There is new research indicating a higher rate of skin cancers in those with sleep apnea. Also there seems to be a link between psoriasis and sleep apnea. Both issues are related to decreased oxygen and blood flow to all the organs, including the skin. Alcohol and obesity can worsen the possibility of sleep apnea. If you find that you are sleepy during the day even after a full night's rest, or if your partner says that you snore, then seem to stop breathing, then snort and breathe again, you may want to get a sleep study done. Having sleep apnea diagnosed and treated has made all the difference for many people in terms of their energy and health, including bringing more blood, oxygen, and nutrients to their skin.

DECREASE STRESS TO DECREASE WRINKLES

A plant can go into shock and shut down from being jerked out of its pot and replanted in another one carelessly or from exposure to extreme temperatures. In the same way, stress is not conducive to your health or your beauty. Many people turn gray or lose hair during high-stress times. If you doubt this take a look at past presidents in their years of office. Scientists estimate, from classic measurements of aging, that the president of the United States ages about two years for every one year in office.

Get a massage. Massage helps stimulate blood flow in the skin, and studies show that it relaxes the entire body, lowering stress hormones after a session. The kneading action may also help break down scar tissue.

Try hypnosis and self-relaxation. Self-hypnosis is a powerful self-soothing skill that can help you relax when you are stressed and also lead you into a peaceful night's sleep.

Get your beauty sleep. If you've ever had a sleepless night, it is pretty easy to so see it on your face, particularly around the eyes. Women who are deprived of sleep have disturbed skin-barrier function, greater water loss from the skin, and highly inflammatory chemicals in circulation. Good sleep also keeps the healthy bacteria in your gut. (Loss of sleep cuts back the amount of healthy bacteria in your intestines. Taking a probiotic when you are not sleeping well is a good preventive measure.) Interestingly, good skin health relies on two brain chemicals that also have antiaging properties: GABA and melatonin. GABA enhances the survival of cells that make collagen. Lack of melatonin may cause thinning of the epidermis. Both of these chemicals are low in people with chronic sleep issues.

Try aromatherapy. In one sleep study where very small amounts of scents from jasmine and lavender were "piped" into a room of sleeping subjects, jasmine outperformed lavender by a long shot in terms of helping people enjoy more restful sleep. People moved less and slept more deeply. The amount used was so small that the subjects could not detect the smell consciously. So a very little dab of this natural scent may prove helpful.

AVOID TOXIC EXPOSURE

Just as a plant can wither from exposure to toxic chemicals in the soil or in the air, so your brain and your skin may wither from both internal and external exposure to toxic substances. On a brain scan, toxic exposure gives the brain a bumpy, scalloped, uneven appearance. This is true of ingested toxins, like alcohol and drugs, as well as environmental toxins, like asbestos and lead paint. Toxic exposure can also leave the skin with bumps, uneven color, rashes, and wrinkling. Here are some things to avoid if you want to protect your skin.

Stop smoking. Nicotine reduces blood flow to the skin, robbing it of that healthy, rosy glow. It also destroys elasticity, which promotes wrinkles. The act of puffing on cigarettes also adds fine lines to the area above your upper lip. Smoking for ten or more years can give you "smoker's face." That's a term Dr. Douglas Model introduced in 1985 when he published a study in the *British Medical Journal* showing that he could identify long-term smokers by doing nothing more than looking at their facial features. The "smoker's face" made the people look older than their true age and included the following characteristics: lines above and below the lips, at the corners of the eyes, on the cheeks, or on the jaw; a gaunt appearance; a grayish tone; and a reddish complexion. More bad news: smokers are three times as likely to develop a certain type of skin cancer called squamous cell carcinoma than nonsmokers, according to a study in the *Journal of Clinical Oncology.*

Avoid glycation end products (AGEs). Eating too many sweets and high-glycemic foods can cause wrinkles. A study in the *British Journal of Dermatology* found that consuming sugar promotes a natural process called glycation, in which sugars attach themselves to proteins to form harmful molecules called advanced AGEs. AGEs damage your brain and also damage collagen and elastin, the protein fibers that help keep skin firm and supple. The more sugar you consume, the more damage to these proteins and the more wrinkles on your face. In addition, cooking with liquids rather than over dry heat reduces the AGEs in your body. Steaming and poaching are better skin-healthy choices than grilling or broiling with dry heat.

Limit caffeine. Too much caffeine from coffee, tea, chocolate, or some herbal preparations dehydrates your skin, which makes it look dry and wrinkled.

Watch the alcohol. Alcohol has a dehydrating effect on the body, sapping moisture from your skin and increasing wrinkles. It also dilates the blood vessels and capillaries in your skin. With excessive drinking, the blood vessels lose their tone and become permanently dilated, giving your face a flush that will not go away. Alcohol also depletes vitamin A, an important antioxidant involved in skin cell regeneration. Alcohol abuse damages the liver and reduces its ability to remove toxins from the body, resulting in increased toxins in the body and skin that make you look older than you really are.

SMILING AND SEX: SKIN-HAPPY HABITS

SMILE! IT DE-AGES YOUR FACE!

A sincere and happy, natural smile may be one of the top age erasers of all. People who have brains that are both calm and focused and who have lives that are healthy in myriad ways tend to be happier and smile more often. Social psychology research has also demonstrated that a person's viewing benefits he or she receives as a gift from God or another person has been correlated with successful aging as well as longevity. So while you are eating your fruits and veggies, drinking water and tea, taking brain and skin healthy supplements, and getting plenty of exercise, another way to give yourself a mini face-lift is to practice the art of daily gratitude, giving yourself something to smile about.

STAYIN' FRISKY OVER FORTY

Frank Sinatra's hit song "You Make Me Feel So Young" could also be translated to say, "You make me *look* so young!" People in loving, flirtatious relationships with regular and happy sex lives look younger. It is true, we all look better when we are in love and enjoying great sex with our partners.

Why? In a way, the brain lives in our skin. Our skin is filled with neuronal connections stemming from the brain, which is why the skin responds to touch so quickly. The skin is a sensual and sexual organ, which is why we can be aroused by caresses. Aging can cause the nerve endings in the skin to lose sensitivity, another reason to invest in your brain-heart-skin-genital health. A healthy brain can help you enjoy a healthy sex life—but did you know that a great sex life can also help your brain, your heart, and your skin? It is one of life's greatest win-wins.

Research from Scotland revealed that making love three times a week can make you look an average of ten years younger! The scientists interviewed thirty-five hundred European and American men and women over a ten-year period on a variety of lifestyle topics. Participants ranged in age from 20 to 104, but most were 45 to 55 years old. The one thing this group had in common was that they looked young for their age, according to a six-judge panel who watched the interviewees through a one-way mirror. The volunteer judges guessed the participants' ages to be from seven to twelve years younger than their actual ages.

A vigorous sex life, according to this study, was the second-most-important determinant of how young a person looked. Only physical activity proved more important than sex in keeping aging at bay. The young-looking participants had sex an average of three times a week. Having sex more frequently than three times a week didn't seem to produce any added benefits. Three was the magic number for keeping that youthful glow.

The study also found out that sex without a good relationship that is very supportive and empathetic doesn't slow the aging process. In fact, cheating and casual sex with different partners may cause premature aging from worry and stress.

A DOZEN LIFE-ENHANCING REASONS TO MAKE LOVE!

Consider the following advantages of a good sex life and you'll wonder why more doctors don't prescribe "Go home and make love to your mate three times a week" to increase happiness, health, and longevity.

- Sex can burn fat, about 200 calories per lovemaking session on average—the equivalent of running vigorously for thirty minutes.
- Sex causes the brain to release endorphins, naturally occurring chemicals that act as painkillers and reduce anxiety as well or better than some antidepressants or antianxiety prescription drugs.

- In men, sex seems to stimulate the release of growth hormones and testosterone, which strengthens bones and muscles and skin tone.
- In both men and women, research has shown that sex also seems to prompt the release of substances that bolster the immune system. Researchers at Wilkes University in Wilkes-Barre, Pennsylvania, found that having sex once or twice a week boosts the immune system by 30 percent.
- A man's orgasm can even be beneficial to women, according to research that indicates that semen can reduce depression in women. One study found that women whose male partners did not use condoms were less subject to depression than those whose partners did. One theory put forth was that prostaglandin, a hormone found in semen, might be absorbed in the female genital tract, thus modulating female hormones.
- Research shows that there are anticancer-producing properties to making love, probably due to positive hormones produced during and after sex.
- People who have lots of sex tend to eat better and exercise more, though which comes first, the sex or the health habits, is unclear. It may be another win-win of life, where one activity naturally boosts the other.
- Having great sex—and a lot of it—can boost the levels of hormones, such as estrogen and DHEA (dehydroepiandrosterone), both of which promote smoother, tighter skin.
- Sex makes you happier than lots of money. According to a study by the National Bureau of Economic Research, a marriage that included regular sex was figured to bring the same level of happiness as earning an extra hundred thousand dollars a year.
- A study at Queen's University in Belfast, Northern Ireland, found that men who have sex three or more times a week can cut their risk of heart attack in half. Regular sex also cuts the risk of men having a stroke in half.

- Endocrinologists at Columbia University found that women who have sex at least once a week have more regular menstrual cycles than those who don't.
- In a study done at Duke University, researchers followed 252 people over twenty-five years to determine the lifestyle factors important in influencing life span. Sexual frequency and past and present enjoyment of intercourse were three of the factors studied. For men, frequency of intercourse was a significant predictor of longevity. While frequency of intercourse was not predictive of longevity for women, those who reported past enjoyment of intercourse had greater longevity. This study suggested a positive association between sexual intercourse, pleasure, and longevity.
- Muscular contractions during intercourse work the pelvis, thighs, buttocks, arms, neck, and thorax. *Men's Health* magazine has gone so far as to call the bed the single greatest piece of exercise equipment ever invented.

SEX KEEPS YOU YOUNG

In August 1982, during my internship year on the surgical floor at the Walter Reed Army Medical Center in Washington, D.C., Jesse was discharged from the hospital. He had been admitted for an emergency hernia operation two weeks earlier and there had been some minor complications. I remember Jesse so vividly now because he was one hundred years old but talked and acted like a man thirty years younger. Mentally, he seemed every bit as sharp as any patient I had talked to that year (or since). He and I developed a special bond, because unlike the surgery interns who spent a maximum of five minutes in his room each day, I spent hours over the course of his hospitalization talking to him about his life. The other interns were excited to learn about the latest operating techniques. I was interested in Jesse's story and I wanted to know about Jesse's secrets for longevity and happiness.

Jesse had his hundredth birthday in the hospital and it was quite an event. His wife, actually his second one, who was three decades younger, planned the event with the nursing staff. There was great love, playfulness, and physical affection between Jesse and his wife. Clearly, they still had the "hots" for each other.

Just before his discharge from the hospital he saw me at the nurses' station writing notes. He enthusiastically waved me over to his room. His bags were packed and he was dressed in a brown suit, white shirt, and a blue beret. He looked deeply into my eyes as he quietly asked me, "How long, Doc?"

"How long what?" I answered.

"How long before I can make love to my wife?"

I paused and he continued in a hushed voice, "You want to know the secret to live to one hundred, Doc? Never miss an opportunity to make love to your wife. How long should I wait?"

A slow smile came over my face. "I think a week or so and you should be fine. Be gentle at first." Then I gave him a hug and said, "Thank you. You have given me hope for many years to come."

Science finally caught up to Jesse thirty years later. Now there is a wealth of research connecting healthy sexual activity to longevity. The lesson from Jesse still rings true today. While there are many ingredients to a long life as we have been exploring, frequent sexual activity with your beloved is one of them too.

BEAUTY AND THE BRAIN

If you apply the ideas in this chapter for six months to a year, you, like Joni, may not need that face-lift after all. And better yet, you'll increase your chances for longevity and decrease your chances of depression, dementia, heart disease, cancer, and obesity. The two biggest organs in your body, your brain and your skin, will be better and you will be healthier, happier, and sexier all at the same time.

CHANGE YOUR BRAIN NOW: TWENTY BRAIN TIPS FOR BETTER BLOOD FLOW, SKIN HEALTH, AND SEX LIFE

1. The brain-skin connection is so strong that some people call skin "the brain on the outside." The health of your skin is a visual indicator of the health of your brain. When you take care of your brain, you will also see improvement in your skin. Don't get that face-lift before you see what brain-smart living can do for your appearance!

2. Even a ten-pound weight loss—if you are losing weight on highly nutritious foods—can decrease the age of your face. When people lose weight on a healthy regimen, others often comment on how much younger they look as well. Those who lose weight by eating low-calorie, low-nutrient foods are more likely to look drawn, wrinkled, and weak.

3. Beauty begins in your brain. An emotionally balanced brain relaxes the facial muscles and produces a more serene appearance and smile. Nourishing your brain also nourishes your skin and makes it more attractive.

4. Exercise beautifies the skin. It increases blood flow and releases toxins, tones your muscles, gives your skin healthy color, and eases wrinkle-producing stress. Make daily exercise a part of your skin-healthy regimen.

5. Drinking water is helpful because it flushes toxins from the body, but to keep your skin moist and help protect it from sun damage make sure you also have at least 1 g fish oil a day. Borage, evening primrose, or black-currant seed oils improve dry skin as well.

6. For healthy skin that glows from the inside out, drink tea: white, green, or black. Tea has anti-inflammatory properties, which may help prevent skin cancer and reverse aging. White tea has been shown to protect elastin and collagen, which help the skin's strength and elasticity, diminishing wrinkles and sagging.

7. People in cultures where high amounts of dark-colored vegetables, seafood, tea, dark fruits, and berries are consumed have younger-looking skin. Those who eat lots of processed meats, saturated fats, white sugar, and starch age faster. Your skin is what you eat, so feed it beautiful ingredients.

8. Ditch the skin-harming suntan and give your skin a naturally golden glow by eating more yellow and orange fruits and vegetables, loaded with healthy carotenoids.

9. Massage helps stimulate blood flow in the skin, and studies show that it relaxes the entire body, lowering stress hormones after a session. The kneading action may also help break down scar tissue. So go ahead and enjoy a spa day now and then, and pamper yourself with a therapeutic massage.

10. You really do need your beauty sleep. Women who are deprived of sleep have disturbed skin-barrier function, greater water loss from the skin, and highly inflammatory chemicals in circulation. Interestingly, good skin health relies on two brain chemicals that also have antiaging properties: GABA and melatonin. Try taking these supplements before bed, to help both your sleep and your skin.

11. Toxins age skin. Avoid smoking, sugar, environmental toxins, grilling foods over high dry heat, and drinking too much alcohol and caffeine to keep your skin toxin-free.

12. Though what you put in your body probably affects your skin's health the most, what you put on your skin also matters. Wash, exfoliate, tone, and moisturize your skin with high-quality skin care products that are free of toxins and contain ingredients that are also good for your brain, such as pomegranate and white tea.

13. Be a grateful person and you'll not only be happier, but you will also smile more, which will help you be beautiful from the inside out.

14. Have you ever felt hungry to be touched? We give and receive feelings of love and security through our skin through hugs and loving touches. Those who are physically affectionate with their friends, spouses, and kids will reap the benefits of regular touch for both their bodies and their brains.

15. One study showed that a loving, vigorous sex life can make people appear seven to twelve years younger. Only physical activity proved more important than sex in keeping aging at bay. If you are in a loving relationship, making love three times a week will make you both look younger than your years.

16. People who have lots of sex tend to eat better and exercise more, though which comes first, the sex or the health habits, we are not sure. It may be another win-win of life, in which one activity naturally boosts the other.

17. Having great sex—and a lot of it—can boost the levels of hormones, such as estrogen and DHEA, both of which promote smoother, tighter skin.

18. Your bed may be the best piece of workout equipment in your house. Sex can burn fat, about 200 calories per lovemaking session

on average—the equivalent of running vigorously for thirty minutes. Your reward for participating in this pleasurable activity is a better-toned body and lots of feel-good and skin-friendly hormones.

19. Anxiety and depression can quickly age a person's body and face. However, exercise causes the brain to release endorphins, naturally occurring chemicals that act as painkillers and reduce anxiety as well or better than some antidepressants or antianxiety prescription drugs.

20. In men, sex stimulates the release of growth hormones and testosterone, which strengthen bones and muscles and skin tone. Regular sex helps keep a woman's antiaging hormones circulating, thus beautifying her brain, body, and skin. If you are tempted to say, "I'm too tired" or "I have a headache," think of sex with your partner as not only fun and connecting but also as one of nature's most pleasurable beauty treatments.

7

CHRIS AND SAMMIE

TREAT DEPRESSION, GRIEF, AND STRESS TO ADD YEARS TO YOUR LIFE

There is a fountain of youth: it is your mind, your talents, the creativity you bring to your life and the lives of people you love. When you learn to tap this source, you will truly have defeated age.

—SOPHIA LOREN

ANXIETY ATTACKS

Sammie was a sunny little girl with a wide, happy smile. She was just entering fourth grade when her mother, Chris, took a job outside the

home for the first time in years. Sammie was often overcome with anxiety. She became anxious when her mom had to work a little late, even though friends and family cared for her for the hour or two before Chris could get home. Then, seemingly out of nowhere, Sammie became obsessed with fear about getting sick. "Her anxiety was getting so intense," Chris said, "that we scheduled an appointment with a therapist. It was odd because there was really nothing in our lives that had happened to this point that would cause her to worry like this. No sickness or death in the family or among friends." Chris and her husband, Steve, reassured Sammie that it would be more likely that an "airplane would crash into their house" than that she would get seriously or fatally ill. Still, the inexplicable fears continued.

One Monday, Sammie came home from school describing a "big magnet machine" her class saw on a film in class, which scared her. "I told her it was an MRI machine," said Chris, and calmly explained what it did. "Then we wrote down her fears together on a piece of paper and I asked her, 'What do we do with our fears?' Sammie wadded up the paper and threw it in the fireplace, a visual way to 'toss fears away.'"

The next day, a Tuesday, Sammie took a spill on her Rollerblades. The following night her knee was very swollen, and she was frightened. Chris took her daughter to a clinic the next day. And that is when, to their shock and surprise, an "airplane crashed into their house," so to speak.

THE DREADED DIAGNOSIS

The doctor asked Chris to call her husband for support when he realized the bulge in Sammie's knee was a tumor. Sammie would be diagnosed with osteosarcoma, an insidious type of bone cancer. "Looking back," Chris said, "though we didn't see it as anything more than a phase of childish anxiety at the time, a few scenes flashed through my mind. There was a family outing when Sammie had some shortness of breath. On the

Fourth of July, she'd not wanted to run with the rest of the kids." Sadly, by the time the cancer had showed up on her knee, it had also gone to her lungs. This was followed by chemotherapy, every three weeks for five to seven days, for one very tough year. Along with the chemo came awful nausea, weakness, and hair loss. The doctors were able to save Sammie's cancer-riddled leg, but 70 percent of her leg and knee was now titanium. She also had to have a thoracotomy, which is a painful major surgery that involved an incision in the chest and spreading of the ribs, so the surgeons could get to and remove the tumors in her lungs.

Sammie finished treatment in November of 2008 and was pronounced cancer free, but in January of 2009, she took a fall at the fairgrounds. The cancer had metastasized to her spine. More treatment, spine surgery, more thoracotomies, more agonizing pain, more of every parent's worst nightmares, multiplied. The three-month scan showed the cancer had spread everywhere. In fact, the surgeon said that there were too many tumors to count. Even so, knowing that Sammie's will to live was so strong, Chris and Steve agreed to try more chemo to slow the growth. However, one night, because she was so low in potassium from her many ordeals, Sammie went into cardiac arrest.

"We literally lost her," Chris said, "as her life ebbed away in my arms." Steve, however, knew CPR and managed to get her heart beating again. When she came to, Sammie reported seeing a bright light and thought the family was at Disneyland because it was such a happy place. Her glimpse of heaven? After this experience, there would be no more chemo, no more medical interventions.

THE HARDEST GOOD-BYE

"We had to shift from helping Sammie to live, to preparing our precious child to die," said Chris. Talk about a task that no mother or father can fathom having to do.

Sammie lived seven more months, but these days were filled with horrendous pain, sickness, and depression. Eventually hospice came, and though Sammie had a glimpse of the happiness beyond death during her heart attack, she still fought hard to live to the very end. The last three days of her life were worse than Chris or Steve could have imagined. There were many seizures, and the long hours of watching their little girl fighting in vain for her life were devastatingly difficult. "Sammie wanted to die at home," Chris explained, a catch in her voice, wiping at tears. "We did all we could for her up to the end, when she died as her father held her tight in his arms."

Those last three days, along with the multiple ordeals over the last three and a half years, would be a trauma that would fill Chris's mind with painful recurring memories that invaded her thoughts night and day.

DESCENT INTO DARKNESS

There was a beautiful funeral, a celebration of Sammie's courageous short life, but after Chris said her final good-bye to her daughter, in many ways she said good-bye to herself. "Sammie was our middle child," Chris shared. "She has an older sister, Taylor, and younger brother, Ryan, and of course I did my best as a parent in crisis, but they were often put on the back burner of our lives, which revolved around saving their sister's life and then helping their sister let go of her life. There was little or no thought of myself during those years as I poured every ounce of my strength into Sammie's care."

To try to numb the great loss, Chris began to eat and to drink alcohol. She had a hard time getting out of bed. "I sank into a depression so low that I told myself I would just make it to the anniversary of Sammie's death and then find a way to end my life. I just didn't feel I could go on. I was seeing a good therapist who worked with me every week, and promised me there would be a shift eventually that would help lessen this debilitating grief. He did not encourage medication, telling me, 'There is no medication that can touch this kind of pain.' But there were things I kept from the therapist, even, like my increased use of alcohol."

THE TURNING POINT

Then one day Chris was visiting with a friend of her sister's. "This lady was very fit and had a positive attitude." Chris was short, just 5'1", weighing a little over 200 pounds and walking in a gloom so heavy she thought she'd never smile—and mean it—again. "This friend had a copy of the *Change Your Brain, Change Your Body Food Journal*," Chris explained. "Flipping through it, I thought, 'Okay, I like this. It makes sense to me. I have to start looking for the brighter side of life.' After all, my choices at that point were to take my life, drink myself to death, or end up in rehab. And I was way too proud to go to rehab."

Chris went home, found the book *Change Your Brain, Change Your Body* on the Internet, and downloaded it to read that evening. She read the whole book (not a short book!) in one night. "There was one part of the book, and I can still remember how I felt when I read the list of things that alcohol stops you from feeling, like empathy and compassion for others. I knew I needed to get my feelings of empathy and compassion back for my other children and husband. I needed to find a way to be happy and whole again for their sakes, and my own."

Chris was a thirsty sponge and she absorbed everything written in the book. She said, "I went hard core into the plan. In fact, I did a twenty-eight-day cleanse. I tossed out all the alcohol, ate no processed food, and began taking fish oil and vitamin D."

The change for Chris was fast and remarkable. "Within eight days I didn't care if I never dropped a pound again. I was free! Because I was eating food that actually nourished my cells, the food and alcohol cravings stopped. I got rid of all the diet drinks and colas. I slept through the night for the first time in four years. And for the first time in years, I didn't wake up in a panic."

Chris continued, "I never looked past the trauma with Sammie for the reasons behind my depression and addiction. I blamed everything I felt on the loss and grief. But as I read *Change Your Brain, Change Your Body*, I took a broader inventory of my life and realized I'd struggled with anxiety for a long while. After the first eight days of getting rid of junk and dealing with ANTs (automatic negative thoughts) according to the methods in the book, my anxiety level dropped from a 10 to a 3. I can actually talk myself out of an anxiety attack now by asking myself questions such as, 'Is it really true?" and 'If even a part of it is true, is there anything I can do to change it?' I also use other antianxiety techniques from the book, but honestly, it is usually so calm inside my brain now that I don't need them often. And remember, I only began this program five months ago."

Chris went from never having run before to running four days a week. She advises, "I really recommend running outside if at all possible. I also got myself some support by getting involved in a group called Running for Women where they taught me how to run and walk in intervals. I run 4.5 miles a day, four days a week. Sometimes I run with a bunch of women but sometimes alone. I don't let anything deter me. I go no matter what because I need those endorphins to help me stay balanced! I may not 'love' the feeling of running while I do it, but I see it as an emotional investment that pays huge dividends. It only takes an hour, four times a week, and the payoff is a calmer brain and healthier body."

In five months, Chris is down 35 pounds and four pant sizes, which she sees as a side benefit to getting her emotional life back thanks to nutritional changes and physical activity. "It's not just that I am losing weight. I am changing the *way* I lose weight. I have lost 8 inches off my waist, my neck has slimmed, and my skin is brighter. Though I'd like to lose some more weight, if you saw me today, you'd never think, 'Oh, there's a woman who needs to lose some pounds!' My weight, with the new muscle, has distributed itself well. I used to be a Weight Watchers gal, but I ate a lot of diet food on that plan, and it only increased my cravings. I'd always gain the weight back. On this plan I don't have cravings because I am eating real food that satisfies. I finally look like I'm in the body I've always belonged in, and I am!"

Chris now admits that the depths of her depression had roots and patterns beyond the loss and trauma. She says, "I'm grieving now, but I'm grieving well." She spoke of taking care of herself the way a mom would take care of a child. "Friends tease me about always having healthy snacks in my purse when we are out and about at kids' events. Lord knows there's nothing redeeming in most snack bars. Trust me, I've looked! I just ask them, 'When your kids were little, did you ever leave the house without a diaper bag packed with everything your child might need, including healthy snacks and a sippy cup? You were prepared because you didn't want to end up stuck somewhere with a cranky, crying baby, right? Well, I'm just treating myself as a good mom would treat her child because I don't want to get stuck somewhere without nutritious food or drink and end up cranky and crying.' "

Chris's daughter, Taylor, is now seventeen years old, and much of her teen years were spent without her parents' full focus because of her sister's major illness. Recently Taylor told her mom, "You are a joy to be around again!" Ryan is eleven now and so supportive of his mom as well. He told Chris, "You don't have two necks anymore!" (Boys at eleven are honest, if anything.) He also told her one day that he wasn't going to have dessert for a whole week as a way to support his mom's efforts. Her husband, Steve, has stood by Chris through every moment of this long ordeal and he verbalizes how proud of her he is, that he knows that every morning she wakes up she has to make a choice to live happy and honor her daughter's memory or slack up and go back to that dark place.

"I'm still grieving but I am grieving better, if that makes sense. When we spend time together as a family, going out somewhere fun, I really have to fight the ANTs. My automatic reaction is sadness and thinking about the hole in our family where Sammie used to be. But I take control of my thoughts and remind myself that Sammie is happy and wants us to be happy too. I focus on what we still have, including one another, and the good memories of Sammie, instead of what we don't have."

LIFE IS GOOD AGAIN

Chris has become a crusader for my new program, *The Amen Solution—Thinner, Smarter, Happier with Dr. Daniel Amen!* Her father, mom, and sister are on the plan. Taylor has lost 17 pounds as well. Friends sometimes approach her and say things like, "I can't believe I'm telling you this because what you went through is worse than what I'm going through, but I'm feeling depressed and I can't help but notice the change in you." She then shares my books with them. "I keep a copy of *The Amen Solution* on my kitchen counter now because I refer to it so often for myself or for someone who asks me about what's happened to me the last few months."

When I was speaking at our northern California Clinic, Chris's friend Mo, who is also the director of outcome research at the Amen Clinics, called Chris and encouraged her to come meet me and hear my talk. Chris's ANTs said to her brain, "Oh, I have too much to do, I really can't." But immediately Chris recognized this as the "old Chris" talking and replied, instead, "I'd love to go." I, for one, am so glad she did, as her story deeply and profoundly touched me.

When I first met Chris she started to cry, which almost made me cry. This happens to me a lot. People come up to me, start to cry, and then tell me how our work has changed their life or the life of one of their loved ones. This has always been the reason behind why I do what I do, despite some of the obstacles that have come our way.

Chris shared, "I'm honoring Sammie's memory in a better way by taking care of myself, and everything else seems to be just falling into place. I don't want to live in the realm of 'childhood cancer.' I want to help people enjoy their lives and overcome whatever issues are troubling them. Sammie didn't want to be 'the cancer kid' or poster child for cancer either. She just wanted to be Sammie, herself."

The family sold their home and moved to a new house shortly after Sammie died. In Chris's case, the house was too full of sad memories and a new home represented a new chapter of life. As Chris began to ponder the other changes she wanted to make in her life, she looked for a job that would support her goals. She applied to work at a specialty shoe store where people get into the right running shoes for the shape of their foot. "I have a strong retail background and really wanted to work for this company, but were they going to hire a middle-aged woman in her forties who weighed 202 pounds? In years past, I would have let that thought intimidate me, but my focus was so strong on getting healthy and I'd been through the death of a child, so really, I felt I didn't have anything to lose in applying, and I took a leap."

Chris eventually secured the position and today absolutely loves her job. "The staff is gentle and kind, and we have a great time on the job together. I am stretching and growing my brain cells by learning so much new information."

She continued, "I am creating a life that I desire instead of what was presented to me. I saw quite poignantly that there is much in life over which we have no control. So I'd better take control of what I can. So I am feeding myself well, taking myself for runs, being nice to myself. And this allows me to be the healthy, happy person who can now give joy to others again."

Chris's sister started a "bunny drive," which happens every spring, when children's hospitals are in most need of gifts for sick kids. The charitable project gathers all sorts of new stuffed animals (five thousand in 2011) and gives them to six local hospitals that care for children, making deliveries on Good Friday. When a child gets a hospital bed during this time, a stuffed animal lies on the pillow with a name tag that says, "From Angel Sammie." For more information on how to help with the bunny drive and to read more of Sammie's story, you can visit www.caringbridge.org/visit/sammiehartsfield.

HOW DEPRESSION, GRIEF, AND STRESS STEAL YOUR LIFE

Depression, grief, and chronic stress all take years off your life and make you look, feel, and think older than your age. Depression is one of the greatest problems and killers of our time, affecting fifty million Americans at some point in their lives. Nearly all of us have either suffered from depression or have known someone who has. Two of my best friends had fathers who killed themselves. Depression, all by itself, is a risk factor for Alzheimer's disease, heart disease, cancer, and obesity. When depression accompanies heart disease, people are much more likely to die earlier than those without it. Grief causes your heart to go into abnormal rhythms, which is one of the reasons why people often experience physical chest pain with the loss of a loved one.

Chronic stress is associated with an increased production of certain hormones that cause metabolic changes that put more fat on your belly and diminish the part of the brain that helps get memories into long-term storage. Whenever you experience depression, grief, or chronic stress, think of it as a health emergency. These negative emotional states have a damaging effect on your physical body and have been associated with obesity, cancer, diabetes, heart disease, and dementia. Unless treated aggressively, depression, grief, and chronic stress will rob your ability to live a long, healthy life.

Chris's story illustrates how these negative emotional states can overwhelm and steal your life, and how to work to heal them. When hit with a severe emotional trauma many people self-medicate the pain with toxic food, alcohol, drugs, or sex. The problem, of course, is that short-term solutions only eradicate the pain for a little while and make the situation worse in the long run.

THE BEST TIME TO START HEALING FROM STRESS IS WHEN IT STARTS

Pastor Gerald Sharon lost his wife in December 2010 after a nine-month battle with colon cancer. It was hell for him. Yet his initial reaction to the crisis had been the opposite of Chris's. Gerald had been a senior pastor for many years and had seen many people fall apart during the stress of crises. He knew that if he was going to survive and be able to take care of his wife and be there for his children he had to get healthy. So in the midst of the family emergency, he went to his physician, saw a nutritionist, got on a healthy-eating regimen, exercised, and lost 50 pounds. "I am a stress eater," he told me. "I knew if I was not careful, I would gain 50 more pounds and die in the process as well."

It is common to suffer physical and emotional exhaustion when mourning. Adapting to the loss of someone you love is one of life's most difficult challenges. Your brain is being asked to adapt, sometimes suddenly, to new routines without the presence of your loved one. Grief can leave you thinking you may be going crazy at times. In her best-selling memoir *The Year of Magical Thinking*, the author Joan Didion writes candidly about the sorts of "tricks" her brain would play on her as she adjusted to the sudden loss of her husband after many decades of a happy marriage. "We might expect that we will be prostrate, inconsolable, crazy with loss," she writes. "We do not expect to be literally crazy, cool customers who believe that their husband is about to return and need his shoes." It will comfort you to know that you are not going crazy; but your deep limbic system, where your emotions of connection and bonding are stored, is going through withdrawal from the person you love as bad as that of any drug withdrawal.

Immediately following a major loss, your prefrontal cortex (PFC) may be overwhelmed by the limbic system (feeling and mood center) and temporarily cease to function well. Even creating a to-do list may feel taxing in the days, weeks, and months following loss or during times of high stress. Accept that you may need to radically shorten your to-do list, and give your brain and body vital time to adjust.

You do not have to react to a crisis with self-destructive behavior. You can respond with love—the type of self-love that opts for caring for your brain and body, so that you can deal with the stresses that will inevitably come your way.

NATURAL WAYS TO TREAT DEPRESSION, GRIEF, AND STRESS TO LOOK AND FEEL YOUNGER

Both Gerald and Chris give us guidance on how to naturally treat depression, grief, and stress. In addition to being kind and patient with yourself, it will help immensely if you can put "Take Exceptional Care of My Brain" at the top of your newly shortened and reprioritized to-do list. This one decision will help you cope better with all the other choices and adjustments you must make during a difficult time. Here are some ways to practice good self-care and brain care during stressful or sorrowful seasons:

- Eating proper food to nourish your body is critical during times of stress. Fruits and veggies will reduce inflammation that leads to more pain and illness. Protein will assist a strong transmission of nerve impulses in the brain and is a key fuel for healing and optimizing brain function. One of the amino acids in protein, tyrosine, will increase the levels of two important neurotransmitters, norepinephrine and dopamine. This habit will help your energy levels and you will feel better physically. If you don't feel like eating solid food, a high-quality protein shake can come to your rescue. Sip on it as you can, even though you may not feel hungry at first. Though your appetite might not present during the worst days, your body needs nutrients now more than ever.

- Get physical exercise, which will help you withstand and heal emotional pain. Force yourself to put on those sneakers and walk, preferably outside so you can also get the benefit of sunlight. You may find, like Chris, that pushing past initial reluctance to get out of the house and move will pay big dividends. Listening to music or an audio book may help make the walk more enjoyable. Even if you don't "love" walking while you are doing it, you will notice an improved mood during the day, which will make this investment in your brain health worth the effort. Yoga may also be soothing if you are feeling anxious as well as sad.
- Take healthy supplements. Smart supplementation is important, such as using omega-3s, optimizing your vitamin D level, and using targeted supplements depending on your own type of brain (more on this in a bit).
- Eradicating the ANTs that infest your mind is critical to keeping your mind fertile for long-term growth. (More on this later in the chapter as well.)
- Cry. Did you know that tears of sorrow have been shown to have toxins in them, where tears of happiness do not? This is one reason that you tend to feel better after a good cry when you are sad. So don't postpone or push back the tears. Let the grief waves flow, and wash some of the pain out of your brain and body with a good cleansing cry.

- Set goals. Having a specific goal is critical for positive brain function. After that good cry, you'll probably feel some lifting of sorrow and this is a good time to treat yourself to something you like to do, that is good for you. You need breaks from sadness! Meet an understanding friend for lunch, browse to your heart's content at a bookstore, play or run with your dog. Watch a sitcom or read a light, funny book if you can. Do whatever soothes and offers even a small bit of joy.
- As soon as possible, learn something new. Perhaps take an art class, a gourmet cooking class, learn how to fly fish, or dance the samba. Learning anything new, especially something outside your comfort zone, will stimulate the growth of new cells and neural connections. When Chris pushed herself to apply for and learn a new job, her brain's neurons responded with growth and healing feelings. If the new subject or skill really catches your attention, you may experience periods of "flow" where you feel almost transported away from worry, regret, and sorrow for a time. A fascinating new hobby or interest can give you mini-vacations from grief. One widow who was struggling to find a reason to live woke up one morning and thought, "I'd like to work at a horse rescue ranch." She followed that small desire volunteering at just such a ranch and found the work to be meaningful and healing. As she comforted the animals, they in turn comforted her. It also put her in contact with compassionate people who shared her love of animals and nature.
- Hydrate. Water is crucial in brain function and will also keep your lymph system working to remove toxins from the immune cells and reduce the possibilities of infections when under stress.
- Make every effort to focus on what you can do to strive to be more loving and grateful. "In my thirty years of helping the bereaved," says Dr. Louis LaGrand, a grief counselor and author of *Love Lives On: Learning from the Extraordinary Experiences of the Bereaved*,

writes, "I am convinced that love is the single most effective coping strategy that will get you through any loss. For the brain, it will minimize, often eliminate, the power of negative thoughts to generate excessive and unnecessary emotional and physical pain." In addition to the great advice of "loving more," I would add, "purpose to be more grateful." The emotions of love and gratitude literally block many negative emotions in the brain. It is hard to really feel grateful and negative or stressed at the same time.

COMPLICATED GRIEF

Most of us will experience the pain grief brings at some point in our lives, either through the death of a loved one, unwanted divorce, or other rejection or loss. In time, most people move on. Though we never forget someone we loved deeply, as months and years pass, the memories of them are less attached to pain and more attached to warm and positive memories. This time varies for all of us, as human beings don't grieve on a set schedule, but there is a syndrome that may perpetuate the grief process longer in some people than others.

In a UCLA study on bereavement, scientists proposed that in addition to normal grief, there is a syndrome called complicated grief. This is a sort of unrelenting grief that never allows for healing and moving on. It activates neurons in the reward centers of the brain (as well as the pain centers), possibly giving past memories addiction-like properties. After the loved one dies, those who adapt to the loss eventually stop getting this neural reward, but those who don't adapt continue to crave it. This is not something people consciously do; it simply happens to some people (10–15 percent of the grieving population) and not to others.

Complicated grief can be debilitating, involving recurrent pangs of painful emotions, including intense yearning, longing, and searching for the deceased and a preoccupation with thoughts of the loved one. Some of these experiences are common to the newly grieving as well, as the brain strives to adjust to new realities. But if the brain *doesn't* adjust, if grief is just as intense years later, additional help may be needed to process through grief and find a meaningful life again. Besides the "natural helps for grief and stress" listed above, someone whose grief has become complicated or has turned into a clinical depression may need therapy, supplements, and medication to move forward again.

Below is a chart that might help you determine if you or someone you care about is simply going through the normal grief process or may have complicated grief or clinical depression.

CHARACTERISTICS OF GRIEF OR DEPRESSION: A SIDE-BY-SIDE COMPARISON

Grief	Complicated Grief or Depression
A normal response to loss that causes distress	Generalized distress—loss of interest, pleasure
May experience some physical symptoms of distress	Physical distress, hopelessness, guilt
Still able to look toward the future	No sense of a positive future
Passive wish for death	Suicidal ideation not uncommon
Retains capacity for pleasure	Change in capacity to enjoy life or things that were formerly pleasurable Persistent flat affect, negative self-image
Still able to express feelings and humor	Bored, lack of interest and expression
Comes in waves	Constant, unremitting
Can cope with distress on own or with supportive listening Medication for grief is an exception, not the rule	May require combined therapy, supplements or medication

KNOW YOUR BRAIN TYPE

Knowing the type of brain you have is critical to getting the right help. When I first started our brain imaging work at the Amen Clinics in 1991 I was looking for the *one* pattern that was associated with depression, ADHD, or bipolar disorder. But our brain imaging work clearly taught us that there was not one brain pattern associated with any of these illnesses. They all had multiple types. Of course, I then realized that there will never be just one pattern for depression, because not all depressed people are the same. Some are withdrawn, others are angry, and still others are anxious or obsessive. The scans helped me understand the type of depression, ADHD, bipolar disorder, overeating, or addiction a person had, so that I could better target their treatment.

This one idea led to a dramatic breakthrough in my own personal effectiveness with patients and it opened up a new world of understanding and hope for the tens of thousands of people who have come to the Amen Clinics and the millions of people who have read my books.

Basically, we see these eight brain types:

1. Impulsive
2. Compulsive
3. Impulsive-compulsive
4. Sad or moody
5. Anxious
6. Temporal lobe
7. Toxic
8. Post-traumatic stress

We see these in all sorts of combinations as well. This is exactly the reason why most psychological and psychiatric treatment programs do not work on a consistent basis. They take a one-size-fits-all approach to illnesses, such as depression, which judging from our brain imaging work makes absolutely no sense at all.

There is much more information, including a detailed self-test, on our website (www.theamensolution.com) and in my other books, such as *Change Your Brain, Change Your Life, Healing ADD, Healing Anxiety and Depression, The Amen Solution*, and *Unchain Your Brain.*

BRAIN TYPE 1: THE IMPULSIVE BRAIN

People with this brain type have low activity in the PFC. Think of the PFC as the brain's brake. It stops us from saying stupid things or making bad decisions. Subsequently, people with this type of brain struggle with impulse control, attention, and disorganization. They have trouble thinking about the consequences of their behavior before they act, which can get them into all sorts of hot water with their health, relationships, work, and money.

One of my best friends is a perfect example of this type. He is on a diet every single day of his life. He wakes up every morning committed to the idea of eating right. He maintains the thought as he passes the first doughnut shop. He starts to sweat as he passes the second doughnut shop. By the third one, he has no willpower left. After completely giving up on his plans by noon, he utters the famous words of all impulsive overeaters: "I'll start my diet tomorrow."

This type is common among people who have ADHD, which has been associated with low dopamine levels in the brain. Research suggests that having untreated ADHD nearly doubles the risk for being overweight and having other medical problems. Without proper treatment, it is nearly impossible for these people to be consistent with any health plan.

My research team and I have published several studies showing that when people with ADHD try to concentrate, they actually get less activity in the PFC, which will cause them to have even less control over their own behavior. For these people, literally, the harder they try to lose weight, the worse it gets.

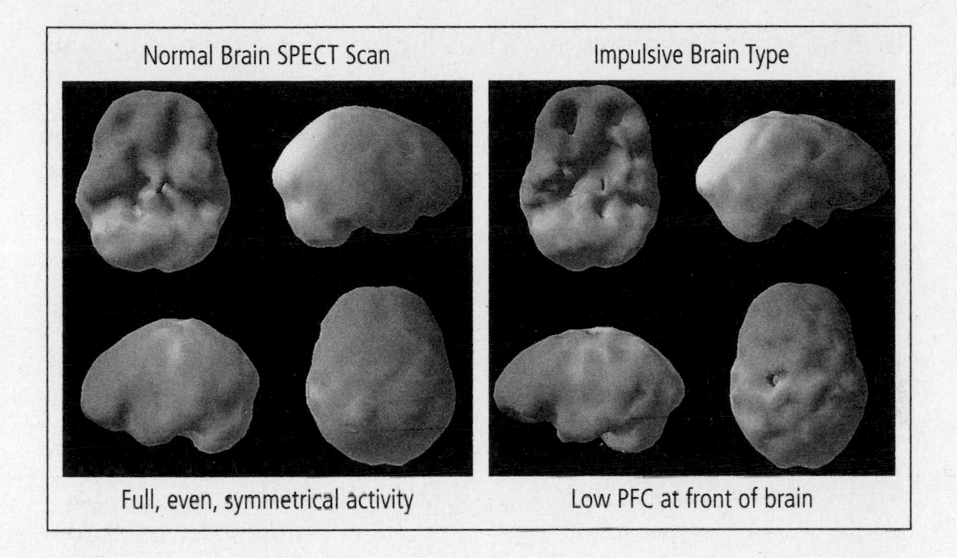

Full, even, symmetrical activity   Low PFC at front of brain

This type is helped by boosting dopamine levels in the brain to strengthen the PFC. Higher-protein, lower-carbohydrate diets tend to help, as do exercise and certain stimulating medications or supplements, such as green tea or L-tyrosine. Any supplement or medicine that calms the brain, such as 5-HTP (5-hydroxytryptophan), typically makes this type worse, because it can lower both your worries and your impulse control.

BRAIN TYPE 2: THE COMPULSIVE BRAIN

People with this type tend to get stuck on negative thoughts or negative behaviors. They often say that they have no control over their behavior and tend to worry, hold grudges, be rigid and inflexible, argumentative or oppositional. The main problem is that they have trouble shifting their attention, so they get stuck on bad thoughts and behavior.

I am often asked, what is the difference between people who are impulsive versus those who are compulsive? Impulsivity is when you get a thought in your head and you just act on it without thinking. Compulsivity is when you get a thought in your head and you feel as though you have to act on it.

The compulsive brain on SPECT scans usually shows too much activity in a deep part of the frontal lobes called the anterior cingulate gyrus. I think of this part as the brain's gear shifter that helps us go from thought to thought or idea to idea. When it functions optimally, people tend to be flexible, adaptable, and go with the flow. When this part of the brain works too hard, usually owing to a deficit in the neurotransmitter serotonin, people tend to be rigid, inflexible, and get stuck on bad thoughts or bad behaviors.

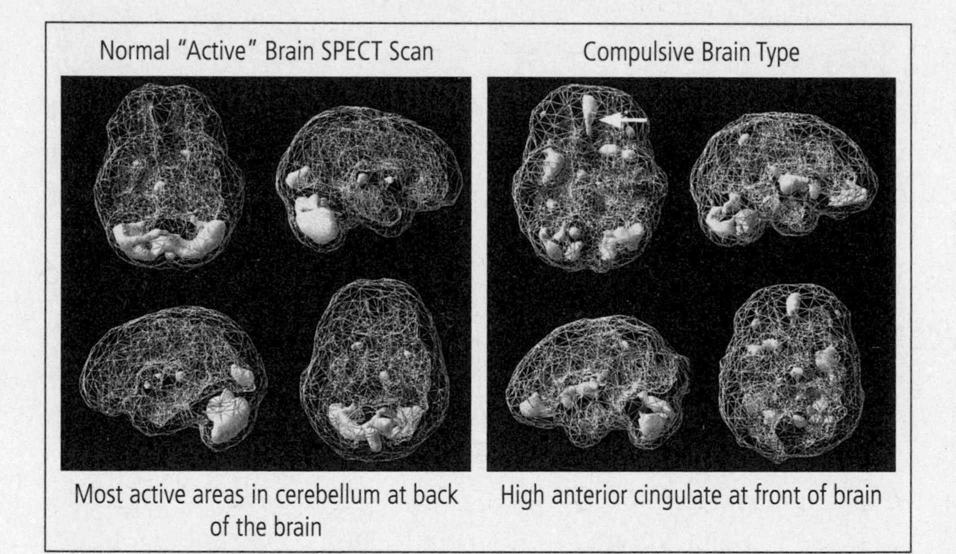

Most active areas in cerebellum at back of the brain

High anterior cingulate at front of brain

Caffeine and diet pills usually make this brain type worse, because these brains do not need more stimulation. People with this type often feel as though they need a glass of wine at night, or two or three, to calm their worries.

Compulsive brain types do best when we find natural ways to increase serotonin. Serotonin is calming to the brain. Physical exercise boosts serotonin as does using certain supplements, such as 5-HTP or St. John's Wort. 5-HTP actually has good scientific evidence that it can be helpful for depression, anxiety, and weight loss.

BRAIN TYPE 3: THE IMPULSIVE-COMPULSIVE BRAIN

On the surface it seems almost contradictory. How can you be both impulsive and compulsive at the same time? Think of compulsive gamblers. These are people who are compulsively driven to gamble and yet have very little control over their impulses. It is the same with this brain type. Our scans tend to show too much activity in the brain's gear shifter (anterior cingulate) region of the brain, so people overthink and get stuck on negative thoughts, but they also have too little activity in the PFC so they have trouble supervising their own behavior.

Barb struggled with oppositional and impulsive behavior as a teenager and still at age forty-eight got stuck on negative thoughts and had trouble controlling her impulses, especially in raising her own teenagers. Many people in her family struggled with alcohol and other addictions, which is very common with this brain type. Barb had tried a number of treatment programs before coming to the Amen Clinics without any success. She had tried stimulants for ADHD, which made her angry, and antidepressants that boosted serotonin, such as Prozac, Zoloft, and Lexapro, which seemed to make her more impulsive. After listening to her story and seeing her scans it was clear she had an impulsive-compulsive brain.

People with this type benefit from treatments that increase both serotonin and dopamine, such as exercise with a combination of supplements like 5-HTP (to boost serotonin) and green tea (to boost dopamine) or medications to do the same thing, such as a stimulant plus a serotonin-enhancing antidepressant at the same time. For Barb, this combination of supplements helped to balance her brain so she could feel emotionally stable. Giving her 5-HTP or green tea alone would have made her worse!

BRAIN TYPE 4: THE SAD OR MOODY BRAIN

People with this type often struggle with depression, negativity, low energy, low self-esteem, and pain symptoms. On brain SPECT scans we often see too much activity, deep within the limbic or emotional part of the brain. With this type, when there are the external factors of stress or grief, the vulnerability is often depression. We often hear of depression running in families or being brought on by stressful early life events.

Gary struggled with chronic sadness and negativity. He remembers feeling sad as a child, which only worsened after he lost his grandfather at age thirteen. At age fifty-seven he felt older than his peers and suffered with arthritis. He had tried psychotherapy and several antidepressants with little effect before he came to see us. His SPECT scan showed too much activity in the limbic or emotional part of his brain, which is commonly seen in mood disorders. For this type, we have had good success by encouraging our patients to exercise, take high-dose fish oil (6 g), and certain supplements, such as SAMe (S-adenosylmethionine), to help improve mood, energy, and pain.

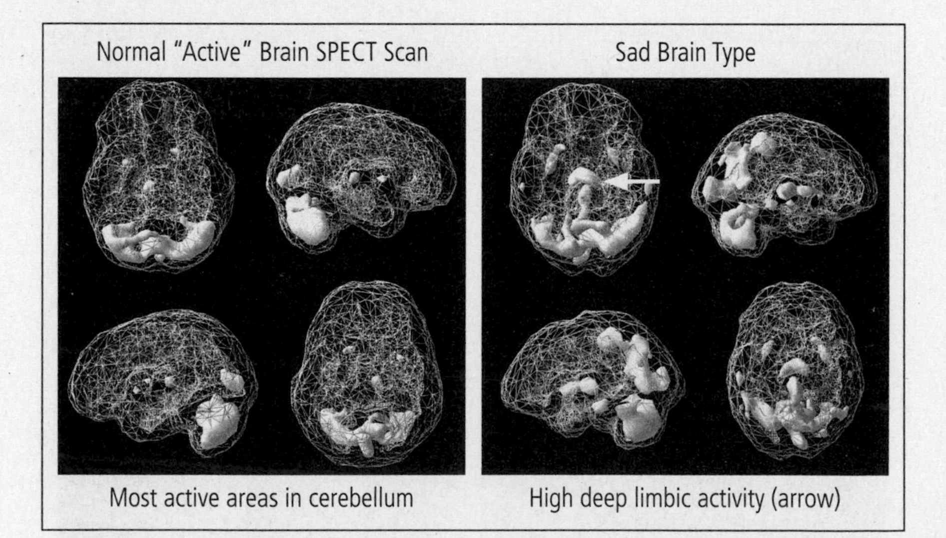

Most active areas in cerebellum

High deep limbic activity (arrow)

SAMe is a nutritional supplement that has good research demonstrating its effectiveness for both depression and pain. Of note, there is a clear connection with physical pain and depression, which SAMe appears to help, as does the antidepressant Cymbalta (duloxetine). When the limbic brain is combined with the compulsive brain, serotonin interventions seem to be the most effective.

BRAIN TYPE 5: THE ANXIOUS BRAIN

People with this type struggle with feelings of anxiety or nervousness. They often feel tense, panicky, and stressed and tend to predict the worst. They are usually conflict avoidant and live with a feeling of angst and that something bad will happen. We often see too much activity in an area deep in the brain called the basal ganglia.

Doreen felt anxious most of the time. She was always waiting for something bad to happen, and she frequently suffered from headaches and stomach problems. Marijuana helped relax her, but it also gave her memory problems. She had tried antianxiety medications but quickly had the feeling that she was becoming dependent on them, a feeling she hated. Doreen's SPECT study showed too much activity in the basal ganglia. This part of the brain is involved in setting a person's anxiety level. When there is too much activity here, often owing to low levels of a chemical called GABA, people often have anxiety and a lot of physical tension.

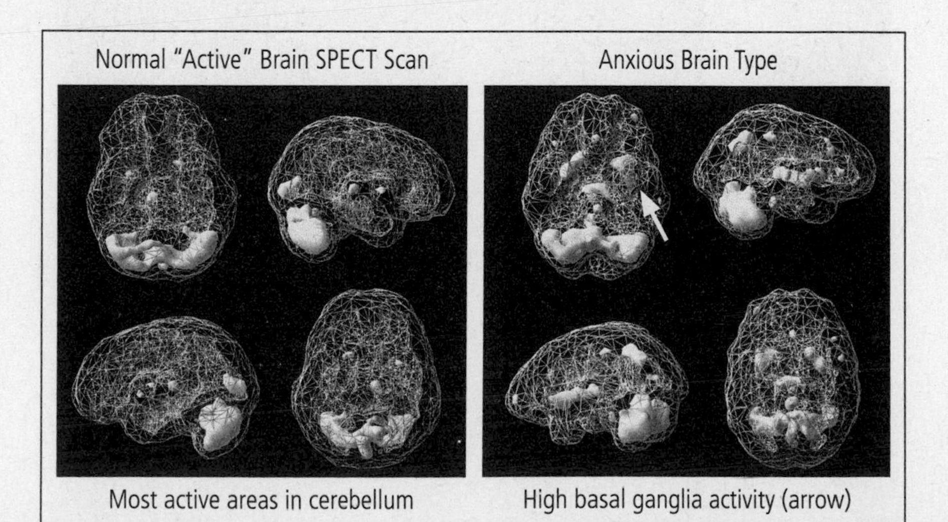

Most active areas in cerebellum — High basal ganglia activity (arrow)

By soothing Doreen's brain with meditation and hypnosis, plus using a combination of B_6, magnesium, and GABA, she felt calmer and more relaxed and noticed a big boost in her energy.

BRAIN TYPE 6: THE TEMPORAL LOBE BRAIN

The temporal lobes, underneath your temples and behind your eyes, are involved with memory, learning, processing emotions, language (hearing and reading), reading social cues, mood stability, and temper control. Trouble in the temporal lobes, often owing to a prior brain injury, can lead to memory problems, learning difficulties, trouble finding the right words in conversation, trouble reading social cues, mood instability, and temper problems. Temporal lobe problems are very common in resistant depression.

Beth, twenty-five, came to see us after her fourth suicide attempt. She had problems with depression and her temper since she was a child. Her moods would fluctuate wildly and she could never predict how she would feel. When she was three years old, she fell down a flight of stairs and was unconscious, but only for a brief period of time. Beth had been on many different antidepressants without success. Beth's brain

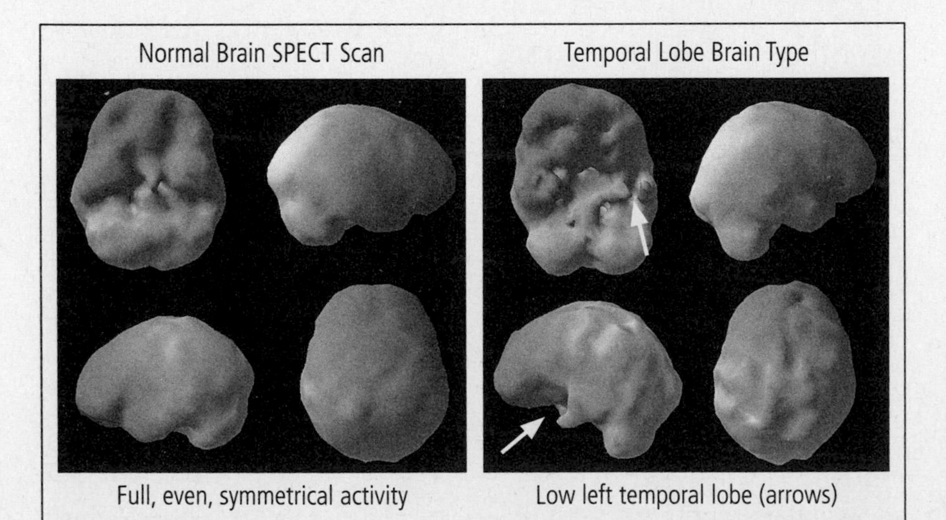

Full, even, symmetrical activity — Low left temporal lobe (arrows)

SPECT scan showed clear problems in her left temporal lobe. For many years now, we have seen trouble in this part of the brain be associated with dark, evil, awful, destructive thoughts, including both suicidal and homicidal thoughts. None of the medications she had been put on were specifically to help stabilize her temporal lobes.

I have found antiseizure medications to be particularly helpful for this brain type. In addition, balancing blood sugar, by eating small meals four to five times a day, making sure to get good sleep, and eliminating sugar are also very helpful. With this combination of treatment, Beth's moods stabilized and she was able to stay out of the hospital and restart college.

BRAIN TYPE 7: THE TOXIC BRAIN

In this brain type we see overall low activity in the brain. There are many potential causes of this brain type, including:

- Drug or alcohol abuse
- Environmental toxins, such as mold, paint, or solvents
- Past chemotherapy or radiation
- Brain infections, such as meningitis or encephalitis
- Lack of oxygen, such as with a strangulation, near drowning, or sleep apnea
- Heavy metal poisoning, such as with lead, iron, or mercury
- Anemia
- Hypothyroidism

Patients with this pattern often feel depressed or sad, have low energy, and are mentally foggy and cognitively impaired.

Will came to see us for resistant depression and brain fog. He had seen six other psychiatrists and had tried numerous medications. He was feeling hopeless, helpless, and worthless. He had frequent suicidal ideas and his family was extremely worried about him. He looked much older than his sixty-three years. His brain SPECT scan showed overall low activity.

This is the classic pattern we see in a toxic brain. I met with him and his wife. He said, and she confirmed, that he did not drink or use drugs. Our first efforts needed to be directed at finding why he had such a toxic-looking brain. After extensive lab and environmental testing, we discovered that he had been working in an office that had extensive mold growth. The office had experienced a flood the year before he first became depressed. Other co-workers also suffered with cognitive impairment.

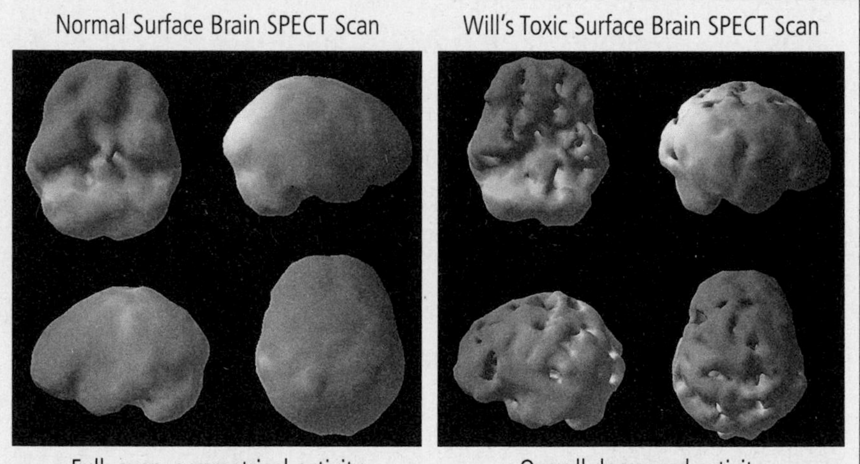

The first step in treating this type of brain is to eliminate the toxin. If a person is drinking or doing drugs, that must stop in order to heal. If there is mold in the environment, that must be fully treated before the person can return to work. If there is severe anemia or hypothyroidism, it is essential to treat it. If there was a loss of oxygen from chemotherapy or radiation, then we know the culprit and can go straight into brain rehabilitation.

BRAIN TYPE 8: THE POST-TRAUMATIC STRESS BRAIN

People who have experienced emotional trauma sometimes develop lifelong stress patterns in the brain, especially if they had vulnerable brains when the traumatic events appeared. On SPECT scans we see that the brain takes on a specific pattern. We call it the diamond-plus pattern, because on scans the pattern appears in the shape of a diamond:

- Increased activity in the anterior cingulate gyrus at the top of the diamond (negative thoughts)
- Increased activity in the deep limbic system at the bottom of the diamond (feelings of sadness)
- Increased activity in the basal ganglia on the two sides of the diamond (anxiety)
- Increased activity on the outside of the right temporal lobe (this is the "plus" part of the diamond-plus pattern), where we think some traumatic memories are stored

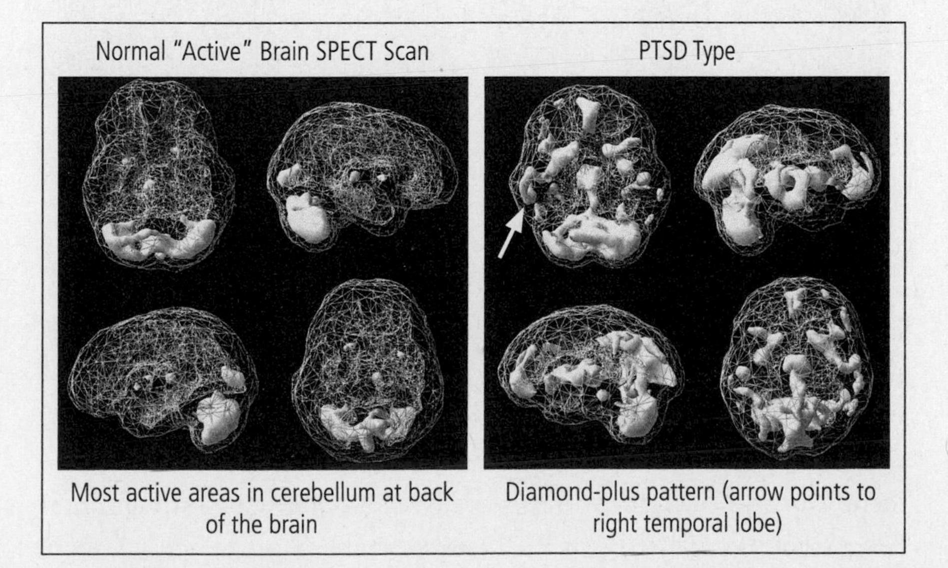

Most active areas in cerebellum at back of the brain

Diamond-plus pattern (arrow points to right temporal lobe)

The scans show a pattern where it looks as though the trauma or traumas get stuck in the brain.

Frank, sixty-six, was the CEO of a large company. He came to see us because his memory was poor and he struggled with anxiety, depression, and excessive alcohol use. His wife had given him an ultimatum to get help or she was going to divorce him. Frank denied having any emotional trauma in his history. His scan displayed the diamond-plus pattern. All of the overactivity in his brain helped us understand why he drank so much. He was trying to put out the fire in his brain that made him feel terrible. But the alcohol made him irritable, and his wife was nearly finished with the chronic stress he gave her. Seeing the diamond pattern led me to ask more pointed questions about past trauma. Again, he denied it. Being the persistent soul that I am, I asked him repeatedly. He kept saying no. When I brought his wife into my office, I asked her. She looked at her husband and said, "Frank has a bad relationship with his father."

"Why is that?" I asked Frank.

"When I initially made a lot of money, I bought a home for my mother but not my father, and he has held a grudge against me ever since."

"Why did you just do it for your mom?" I asked.

"When I was growing up in the poor part of Chicago, my parents were separated and I was raised by my mom. My dad was not around. But my mom was a drug addict and often not home. When I was a young adult, she got clean and I wanted to help support her getting well."

"And you have no emotional trauma in your past?" I said, wondering how he could forget the years of being raised without a father by a drug-abusing mother. My, how the brain can block out pain; it is indeed a sneaky organ.

It was at that moment that Frank's face changed and he started to sob. He had forgotten all of the times when his mother did not come home, when he was left alone, when he thought she was dead, or when she brought over men who were very scary to him. Frank was loaded with trauma he had never processed, and subsequently the trauma still lived, wreaking havoc, blocked in his emotional brain. His overactive brain caused him emotional pain and he used alcohol to try to put the brain fires out. Of course, the alcohol brought him all sorts of other problems and made him more distant from those he loved.

To calm the diamond-plus pattern and eradicate the past emotional traumas seen on scans, we often refer people to a special psychological treatment called EMDR, or eye movement desensitization and reprocessing. I published a study a number of years ago on using EMDR with six police officers who developed post-traumatic reactions after being involved with shootings. At the beginning of the study all of the officers were out of work on stress leaves. After eight to ten sessions their brains were calmer and they all went back to work. You can learn more about EMDR at www.emdria.org. We also have people engage in all the longevity and brain healthy strategies listed in this book.

It is common to have more than one of the eight brain types listed above. If that is true for you, work on the most painful type first and then the others. You can also join our online community at www.theamensolution.com to learn more about your own brain and what to do with combination patterns. Also, when people struggle with resistant problems, a brain SPECT scan can provide additional valuable information.

ERADICATE THE ANTs

One of the techniques that significantly helped Chris heal from her grief was ANT therapy, or learning how to not believe every stupid thought that went through her brain. She learned how to challenge and question the negative thoughts running around her brain. At the Amen Clinics we call it learning how to kill the ANTs. When the ANTs, those negative thoughts that pop up in your brain automatically and seemingly out of nowhere, are left unchecked they steal your happiness, torment you, and can literally make you old, fat, depressed, and feebleminded.

The following exercise to kill ANTs is so simple that you may have trouble believing how powerful it is, but it can change your whole life. Your suffering diminishes and your health and happiness improve. A number of research studies have found this technique to be as effective and as powerful as antidepressant medication for depression.

ANT THERAPY DIRECTIONS

1. Whenever you feel sad, mad, nervous, or out of control, draw two lines vertically down a piece of paper, dividing it into three columns.

2. In the first column write down the ANTs going through your brain.

3. In the second column identify the type of ANT. Therapists typically describe nine different types of ANTs (see table below).

4. In the third column, talk back, correct, and eradicate the ANTs. Were you good at talking back to your parents when you were teenagers? I was excellent. In the same way, you need to learn to be good at talking back to the lies you tell yourself.

ANT	Type of ANT	Eradicate the ANT
I will never be happy again.	Fortune-Telling	I am sad now, but I will feel better soon.
I am a failure.	Labeling	I have succeeded at many things.
It is your fault!	Blame	I need to look at my part in the problem.
I should have done better.	Guilt Beating	I will learn from my mistakes and do better next time.
I am old.	Labeling	I am getting younger every day.

Summary of the Nine Different Types of ANTs

1. Always Thinking: Thoughts that overgeneralize a situation and usually start with words like *always, never, everyone, every time*
2. Focusing on the Negative: This occurs when you only focus on what's going wrong in a situation and ignore everything that could be construed as positive
3. Fortune-Telling: Predicting the future in a negative way
4. Mind Reading: Arbitrarily believing you know what another person thinks, even though she has not told you
5. Thinking with Your Feelings: Believing your negative feelings without ever questioning them
6. Guilt Beatings: Thinking in words like *should, must, ought,* or *have to*
7. Labeling: Attaching a negative label to yourself or others
8. Personalization: Taking innocuous events personally
9. Blame: Blaming other people for the problems in your life

THE WORK: ANOTHER TECHNIQUE

Another ANT-killing technique that I teach all of my patients is called the Work. It was developed by my friend Byron Katie and explained so well in her book *Loving What Is.* Katie, as her friends call her, described her own experience suffering from suicidal depression. She was a young mother, businesswoman, and wife in the high desert of Southern California. She became severely depressed at the age of thirty-three. For ten years, she sank deeper and deeper into self-loathing, rage, and despair, harboring constant thoughts of suicide and paranoia. For the last two years, she was often unable to leave her bedroom and care for herself or her family. Then one morning in 1986, out of nowhere, Katie woke up in a state of amazement, transformed by the realization that when she believed her thoughts, she suffered, but when she questioned her thoughts, she didn't suffer. Katie's great insight is that it is not life or other people that makes us feel depressed, angry, stressed, abandoned, and despairing: It is our own thoughts that make us feel that way. In other words, we live in a hell of our own making, or we live in a heaven of our own making.

Katie developed a simple method of inquiry to question our thoughts. It consists of writing down any of the thoughts that are bothering us or any of the thoughts where we are judging other people, asking ourselves four questions, and then doing a turnaround. The goal is not positive thinking but rather accurate thinking. The four questions are:

1. Is what I'm thinking true?
2. Can I absolutely know that it's true?
3. How do I react when I believe that thought?
4. Who would I be without the thought? Or, put differently, How would I feel if I didn't have the thought?

After you answer the four questions, you then take the original thought and completely turn it around to its opposite and ask yourself whether the opposite of the original thought that is causing your suffering is not true or even truer than the original thought. Then turn the thought around and apply it to yourself and to the other person (if another person is involved in the thought).

Here's an example: Rosemary's husband of thirty-four years died of cancer. Rosemary was the alumni director at my college and we have been friends for many years. After John's death she was very sad and lonely. I helped her work through some of her grief. Two years after John died, Rosemary wanted to start dating again. She loved being in a close relationship. But she told me, "No one would ever want a seventy-five-year-old woman." So we did the Work on that thought. First I asked her this series of questions:

1. Is it true that no one would ever want a seventy-five-year-old woman? "Yes," she said. "I am too old to date."

2. Can you *absolutely* know that it is true that no one would ever want a seventy-five-year-old woman? "No," she said. "Of course I can't know that for sure."

3. How do you feel when you have the thought "No one would ever want a seventy-five-year-old woman"? "I feel sad, hopeless, angry at God, and overwhelmed by my loneliness," she answered.

4. Who would you be or how would you feel if you didn't have the thought "No one would ever want a seventy-five-year-old woman"? "Well, I would feel much happier, more optimistic. I would feel like my usual self," she said.

Then I coaxed her to turn the original thought around. "No one would ever want a seventy-five-year-old woman." What is the opposite? "Someone will want a seventy-five-year-old woman." Okay, so which is truer? "I don't know," she said, "but if I act like no one will want me, then no one will want me." After our exercise, Rosemary started dating again.

A year later she met Jack. When I sat with Rosemary and Jack for the first time it felt like I was with two fifteen-year-olds who had just fallen in love. They were married the next year and will soon celebrate their five-year anniversary.

All of us need a way to correct our thoughts. Just think about what would have happened to Rosemary if she hadn't killed the ANTs that were stealing her happiness and robbing her joy. She would have died a lonely old woman. I have seen these four questions dramatically change people's lives. They can do the same for you.

Rosemary and Jack

ADOPT A POSITIVE ATTITUDE TOWARD AGING

Leroy "Satchel" Paige overcame racial discrimination to break into major-league baseball as a pitcher at the age of forty-two. After a career spanning forty years, he was elected to the Baseball Hall of Fame in 1971. When asked about his prowess at an age when most players are long retired, Paige responded by asking: "How old would you be if you didn't know how old you was?" Excellent question.

Research shows that people with positive, optimistic attitudes about aging outlive those who have a negative, pessimistic view of getting older, by an average of more than seven years.

How powerful is attitude? In the book *Counter Clockwise*, author Ellen Langer writes: "Simply having a positive attitude made far more difference than any to be gained from lowering blood pressure or reducing cholesterol which typically improve life span by about four years. It also beats the benefits of exercise, maintaining proper weight, and not smoking, which are found to add one to three years."

Drowning out ANTs is vital in recovering from loss and the stresses of life, but it can also be a life-lengthening skill when it comes to how you think, specifically, about aging. Visualizing yourself happy, engaged, and healthy as you get older is a wonderful exercise that may add years to your life.

CHANGE YOUR AGE NOW: TWENTY BRAIN TIPS TO HELP YOU THROUGH STRESS, GRIEF, AND DEPRESSION

1. When going through stress, grief, or depression many people are tempted to self-medicate with alcohol, not realizing that alcohol is a depressant! It numbs you from the good, healing feelings of connection and empathy with others that help you traverse grief and pain. Excessive alcohol really only delays and multiplies pain. Say no to alcohol and yes to a healthy healing path.

2. Because loss takes such a hit on our brain and body, it may help you, like it helped Chris and Gerald, to focus on going "hard core" into getting healthy. Getting radically healthy gives your brain and body something positive to throw yourself into and yields positive rewards, emotionally and physically.

3. Long-term caretaking takes a huge toll on the caretaker. Remember to "nurture the nurturer," and learn the art of radical self-care. If you do not replenish yourself while caring for others, you may not be there for them for long. Let others pitch in and take a break to go for a walk, read something inspiring, see a funny movie, or take a guilt-free nap.

4. Take Chris's tip and make sure you don't get stuck somewhere without nutritious food or drink. Almonds or walnuts are a great portable snack to keep in your car, briefcase, or purse. "Treat yourself like

a good parent would treat a child so you don't end up somewhere hungry, thirsty and cranky," especially during stress or loss.

5. Don't let a loss or tragedy leave you a one-dimensional person completely defined by your loss. You are a person who experienced and survived a great sorrow, yes. And you are a person with gifts, talents, and a boatload of compassion to share. Honor the memory of your loved one by taking care of your health and living a full, rich life of benevolence.

6. When you begin to feel anxious or upset you may be dealing with ANTs. Learn to develop an internal anteater to recognize and eliminate these pesky creatures.

7. Question your thoughts. One of the best ways to eliminate the ANTs is to constantly question your negative thoughts. Whenever you feel sad, mad, nervous, or out of control, write down your negative thoughts, ask yourself if they are really true, and start talking back to them. You do not have to believe every thought you have. This exercise can help shift your perspective from negative to positive in seconds and will begin to become automatic with practice.

8. Situational depression is a normal part of grief. Time, tears, reaching out to others, and taking care of yourself will normally ease the pain. But complicated grief or chronic depression that leaves you with suicidal thoughts or unable to cope after many months needs to be treated like a medical emergency. Seek help for this kind of grief immediately. We help many people who are "stuck in grief" through the wonderful resources at our clinics.

9. As Chris pointed out, there is much in life over which we have no control. All the more reason to take charge over what you *can* control: namely your health and happiness, especially after a setback or sorrow. Love yourself and tend to your brain by feeding yourself well, taking yourself out for nice walks or runs, and speaking kind words to yourself.

10. Because grief is draining, you may be tempted to say no to invitations from friends. Try saying yes more often, pushing past the automatic desire to seclude. Being with loving, patient people is healing to a brain overwhelmed by grief. Getting out also gives your brain much-needed "breaks" from overthinking and sadness.

11. The brain can become overfocused on the person you lost, to the exclusion of others who need you. When grief is new, try setting aside a specific time each day to focus on and think about the person you lost: Journal, cry, pray, or whatever you need to do. Then let it go, and turn your mind to others who need you to be fully present for them. Practice the art of "being here now."

12. Don't underestimate the power of changing your diet for the better, adding exercise and a few brain-smart supplements (like vitamin D and fish oil) to quickly turn a low mood around. Feeling better provides its own motivation. Once you experience a lift in mood after tending to your brain and body, you'll be hooked on the feeling.

13. If you have an impulsive brain type, you will be helped by more dopamine. Higher-protein, lower-carbohydrate diets tend to help,

as do exercise and certain stimulating medications or supplements, such as green tea or L-tyrosine. Calming supplements or medications can actually make this brain type worse.

14. If you have a compulsive brain type, you may have a very hard time letting go of painful negative thoughts. Serotonin is calming to this brain type. Physical exercise boosts serotonin, as does using certain supplements, such as 5-HTP or St. John's Wort.

15. If you have an impulsive-compulsive brain type, you may impulsively reach for something unhealthy, then compulsively get "stuck" on doing it over and over. You need to increase both serotonin and dopamine. A combination of exercise with a calming supplement like 5-HTP (to boost serotonin) and green tea (to boost dopamine) can help balance the brain naturally.

16. If you have a sad or moody brain type, you tend to feel depressed, which can also make you feel achy all over and lethargic. For this type, we recommend exercise, a high-dose fish oil (6 g), and certain supplements, such as SAMe, to help improve mood, energy, and pain.

17. If you have an anxious brain type, you may feel tense, nervous, and unsettled inside. Exercise, meditation, hypnosis, and a combination of vitamin B_6, magnesium, and GABA may be helpful.

18. If you have a temporal lobe brain type, you may struggle with memory and learning difficulties, mood instability, dark thoughts, and/or temper problems. Anti-seizure medications, along with balancing blood sugar and getting good sleep, are usually helpful.

19. If you have a toxic brain type, you may suffer with mental fogginess, low energy, and cognitive impairment. Drug or alcohol abuse and environmental toxins are two common causes. Eliminate the toxins and get on a brain healthy program.

20. Boost omega-3s (fish or fish oil) and vitamin D levels to help counteract a low mood.

8

ANTHONY, PATRICK, NANCY, AND MORE ON REVERSING BRAIN DAMAGE

MAKE YOUR BRAIN BETTER, EVEN IF YOU HAVE BEEN BAD TO IT

I have replaced a part of me that had slowly slipped away.

—Fred Dryer

AD: THE NOTRE DAME KILLER

In July 2007, Anthony Davis came to see me as a patient at the Amen Clinics. He was concerned about the cognitive problems he saw in other former retired professional football players.

AD, as he is called by most who know him, is a College Football Hall of Fame running back from the University of Southern California. AD is called the Notre Dame Killer, because in 1972 he scored six touchdowns against the University of Notre Dame. The students at Notre Dame hated AD so much that they put his picture on the walkways of the campus so they could walk all over him. In 1974, he scored four more touchdowns against Notre Dame.

AD had heard about us and thought perhaps we could be helpful to him. At age fifty-four, AD's brain looked like he was eighty-five. It showed clear evidence of brain trauma to the prefrontal cortex and left temporal lobe. For the last twenty years the work at the Amen Clinics has been about brain rehabilitation. We have demonstrated over and over that the brain has the ability to improve after trauma and when we change or improve people's brains, we change their lives. AD's case was no exception. I put AD on a group of brain enhancement supplements that included high-quality fish oil, a comprehensive multiple vitamin and mineral supplement, and supplements targeted to support blood flow and neurotransmitter levels in the brain. Within several months AD told me that he felt better and had more focus, better energy, and stronger memory. I decided to rescan AD in January 2008. His follow-up scan showed significant improvement in blood flow and activity.

Underside Surface View of AD's Brain SPECT Study

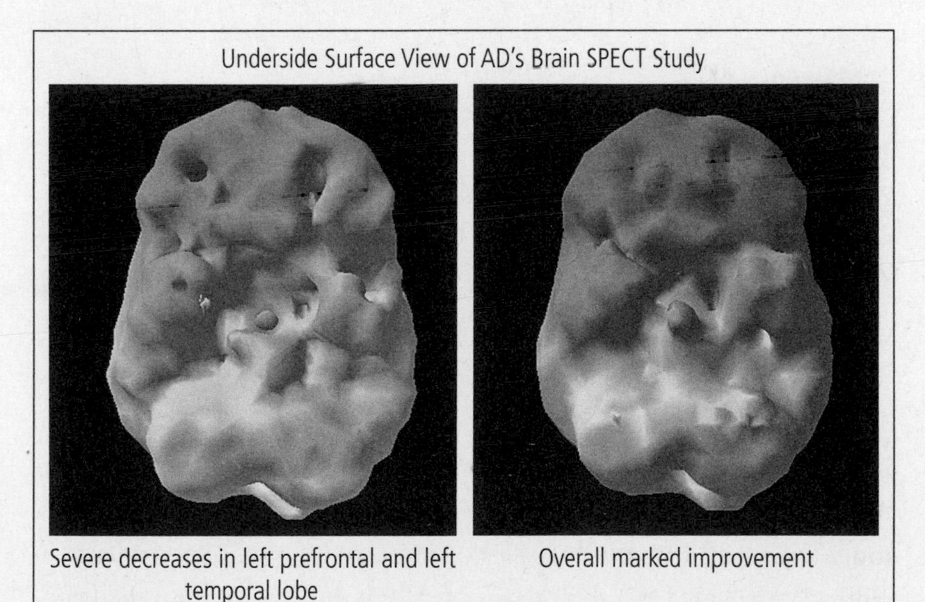

Severe decreases in left prefrontal and left temporal lobe

Overall marked improvement

Through my relationship with Anthony I met many other active and retired NFL players, and he was the impetus for our large-scale study of brain injury and brain rehabilitation in professional football players. At the time, the NFL was still saying it did not know if playing football caused long-term brain damage but had never done the studies to prove it one way or the other. My colleagues and I decided to tackle it. To date, we have scanned and treated 115 active and retired players. Clear evidence of brain damage was seen in almost all players.

Typical NFL SPECT Scan

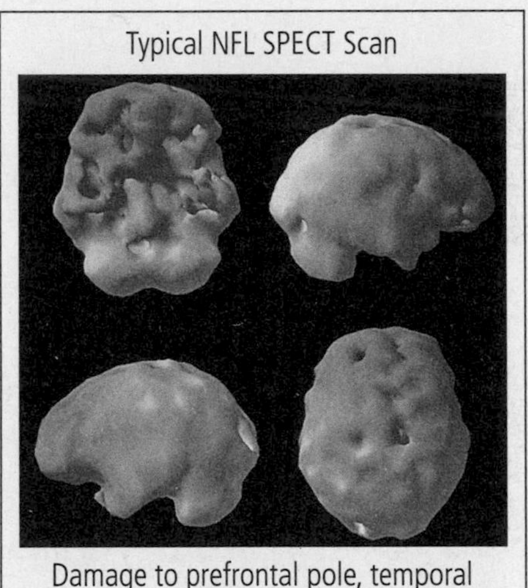

Damage to prefrontal pole, temporal poles, occipital lobes, and cerebellum

The most exciting part of our study we have seen is that recovery and improvement in function is possible, even if the brain damage occurred decades earlier, which was true for most of our players. Seventy percent of our players showed significant improvement on their SPECT scans and their neuropsychological testing. We found after our first five follow-up scans that our initial group of supplements was not powerful enough for the brain damage we saw. That led us to develop a second group of supplements that have made a much more substantial difference, especially our fish oil and Brain and Memory Power Boost formula.

COMPONENTS OF NFL BRAIN REHAB PROGRAM TO REVERSE BRAIN DAMAGE

To help reverse brain damage and facilitate recovery and improved function for NFL players, we've integrated the following components into their rehabilitation:

Education on brain health
- Stop doing things that hurt your brain.
- Start doing things that help your brain.

Optimal nutrition education
Weight-loss group for those who needed it
Coordination exercises
Natural supplements, including fish oil and Brain and Memory Power Boost

For those players who were depressed or demented, we did more. I acted as the psychiatrist for a number of our players or as a consultant to their own physicians. For many, I prescribed natural antidepressants, such as SAMe, because it also helps with pain. If the supplements were not strong enough, I prescribed medications. A number of our players also opted to do hyperbaric oxygen therapy (HBOT), which we have seen improve blood flow to the brain and improve neurofeedback as well. Over the years I have been impressed with the ability of HBOT to increase blood flow to damaged brains. Dr. Paul Harch, one of the world's foremost experts in HBOT, and I did a study of forty soldiers who had experienced brain injuries from IEDs (improvised explosive devises) in Iraq and Afghanistan. We used before-and-after SPECT scans and neuropsychological test data. The results were impressively similar to what we have seen in our NFL players. Neurofeedback uses electrodes first to measure brain electrical activity and then therapists teach patients how to change it. Five examples follow.

ROY

Roy Williams came to see us at age seventy-three. He is part of a three-generation NFL family. He played for the San Francisco 49ers. His son Eric played for the Dallas Cowboys and his grandson Kyle played for the Seattle Seahawks. Roy's cognitive testing scores for attention, reasoning, and memory were in the normal range, but he was significantly overweight at 334 pounds, which was too much for his 6'7" frame. His SPECT scan showed overall decreased brain activity.

When I told Roy about the research studies that report that as your weight goes up the size of your brain goes down, I got his attention. When I added that brain shrinkage is associated with aging, he got the picture and said he wanted to do whatever it took to get a younger brain. Roy runs a highly successful business that helps families transition their wealth to the next generation, so he was not too keen on having a smaller, older brain.

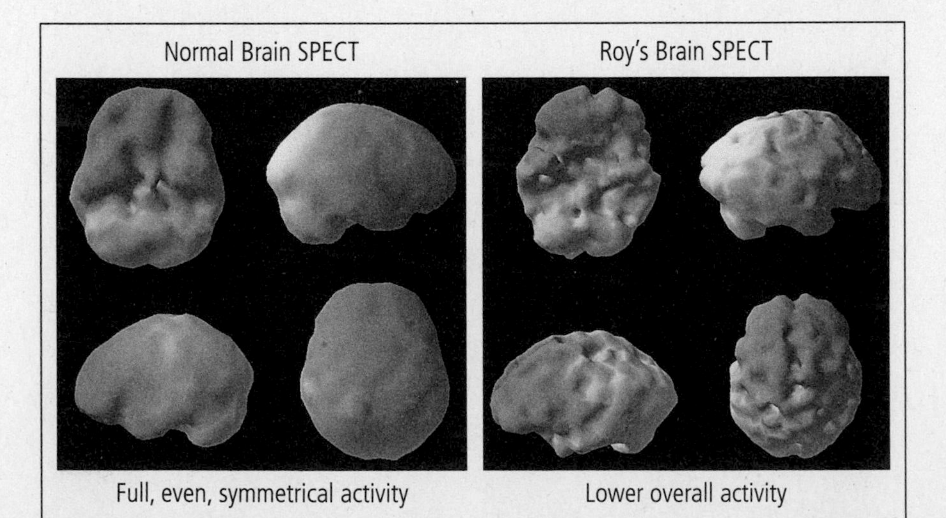

Full, even, symmetrical activity — Lower overall activity

Roy eliminated his bad brain habits and adopted a host of new ones. When he came back for retesting a few months later, he had lost 30 pounds, but more astounding was the fact that his attention, memory, and reasoning scores had improved. His brain was getting younger! Moreover, his wife says he now has the energy of someone who is forty years old, which initially really irritated her, but now that's changed. Over time our program ideas rubbed off on her and she has lost weight too.

MARVIN

Marvin Fleming is another example of how a severely damaged brain can recover. Marvin is the first player in NFL history to play in five Super Bowls. He played tight end for twelve years for the Green Bay Packers and then the Miami Dolphins, including the Dolphins' perfect season in 1972. He was sixty-seven when he first came to see us and his brain was in trouble. Marvin is one of the nicest people we have had the privilege of helping. He is funny, caring, and always looking for ways to improve himself.

When I asked him if he ever had a brain injury he said no. Brain injuries are a common cause of premature aging and cognitive dysfunction. The brain is very soft, about the consistency of soft butter, and the skull is very hard, with many sharp bony ridges. I thought to myself, "You played tight end for twelve seasons in the NFL, how could you *not* have had a brain injury!" So I pressed him. Marvin seemed so proud of himself because he did not remember ever getting his bell rung, being unconscious, seeing stars, or being confused on the football field, like almost all of our 115 players. But I persisted. I had seen his brain and it showed clear evidence of brain injury. I asked about other potential causes of injuries from childhood, adolescence, and outside of football, such as from motor vehicle accidents, falls, or fights. He persisted in saying no. I have been doing this a long time and had seen thousands of scans like Marvin's, and knew better.

"Okay, Marvin, last time, then I will leave you alone: You are telling me you do not ever remember a car accident, fight, or fall or a time when you played football where you hit your head so hard it caused changes in your awareness or thought process?"

What happened next in my office is so common it is a running joke at the Amen Clinics. Ask patients ten times whether or not they had a brain injury and those who initially say no may actually end up remembering multiple occasions on which they lost consciousness or were in severe car accidents. Our research director, Kristen Willeumier, Ph.D., was in the interview with Marvin and gave me a knowing look.

Marvin's face changed. The right hemisphere of his brain had an "aha" memory experience and it was all over his face. "I am so sorry I lied to you, Dr. Amen. When I was in college at the University of Utah, we were driving from Utah to California in the snow and our car skidded off a mountain road and we fell 150 feet to a riverbed below. I was knocked unconscious and my friends had to drag me out of the car so I wouldn't drown."

I wondered how you forget such an emotionally powerful event. But I have seen it happen so many times in my work. Take all the head hits he had given and received in football, plus the car accident, plus whatever else he did not remember, no wonder his brain looked like it was in deep trouble.

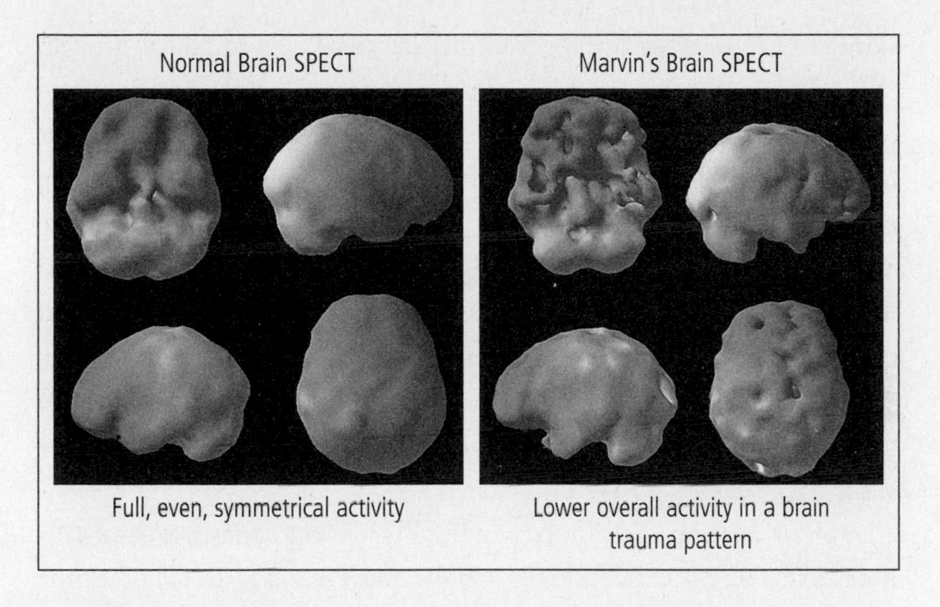

Full, even, symmetrical activity

Lower overall activity in a brain trauma pattern

All of our players also undergo extensive cognitive testing. Marvin's general cognitive testing was not good. What Marvin had going for him was a great personality and a willingness to do the things Dr. Willeumier and I asked.

We asked Marvin to lose weight (he was a "sugaroholic" who would eat frosting out of the can without cake). We also gave him a multivitamin, high-dose fish oil, Brain and Memory Power Boost (our specially

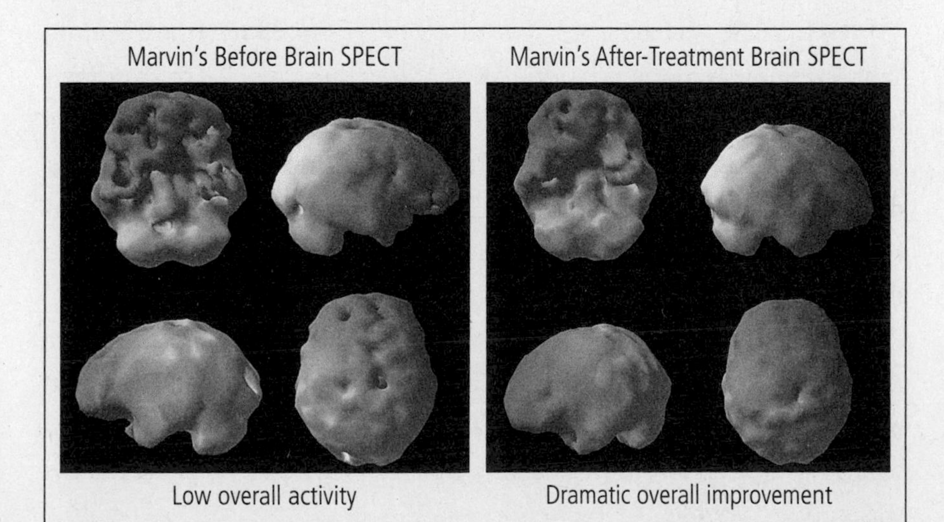

Low overall activity

Dramatic overall improvement

designed supplement geared to enhance brain function), HBOT to boost oxygen to his brain, and we had him increase his exercise.

Two years later, his brain looks dramatically younger, as does he; he has lost 20 pounds, and his cognitive scores have improved by as much as 300 percent.

Typically, the brain becomes less active and less efficient with age. Marvin's brain, like many of the retired NFL players in our study, became more active and more efficient.

FRED

We have dozens of great testimonials and e-mails from our players. One of my favorite ones came from Fred Dryer, the famous Los Angeles Ram defensive tight end who later became an actor and television star of the popular show *Hunter.*

"With the supplements and certainly with the neurofeedback sessions themselves, I have replaced a part of me that had slowly slipped away," he said. "It is very odd to describe the feeling but it is while going through the program, I noticed mental energy and 'speed' of thought and cognition that I 'recognized' I had lost!

"Playing a contact sport for so many years did so much cumulative damage it actually mesmerized me into not noticing the slow progression of brain function loss. It was only when I began to 'feed' my brain with the supplements and at the same time go through the neurofeedback sessions that I began to notice just how far my brain function had slipped. I wish I had knowledge of this science-technology while I was playing professional football. It would have helped prevent all of what had been lost over the years."

CAM

At thirty-four, Cam Cleeland was one of our younger retired NFL players. He played for the New Orleans Saints, New England Patriots, and St. Louis Rams. He volunteered for our study because he was struggling with depression, irritability, frustration, high stress, obsessive thinking, memory problems, and marital problems.

Cam had been diagnosed with a total of eight concussions—three in college and five in the pros. Cam's SPECT scan showed clear brain damage and his Microcog (a test of neuropsychological function) showed significant decreases in general cognitive functioning, information processing speed, attention, memory, and spatial processing.

After eight months on our brain rehabilitation program, Cam reported feeling much better and noticed significant improvements in his attention, mental clarity, memory, mood, motivation, and anxiety level. He felt his anger was under greater control and he was getting along better with his small children.

His SPECT scan showed dramatic improvement in the areas of his temporal lobes (memory and mood stability), prefrontal cortex (attention and judgment), and cerebellum (processing speed). His Microcog showed dramatic improvement as well.

CAPTAIN PATRICK CAFFREY

While deployed in Afghanistan in 2008 Captain Patrick Caffrey, a combat engineer officer, was in the middle of phasing in new, specially armored vehicles. These are the vehicles all of our troops ride today. "We knew one thing about them," he said. "They could take an enormous blast and you'd be able to walk away, unscathed—or so we thought."

One of the many tasks of the Second Battalion, Seventh Marines Combat Engineer Platoon was to conduct route clearance—the intense mission of taking mine detectors and other special detection equipment on roads laden with mines and IEDs. His mission was to find and clear them from the road so that logistics, convoys, and the infantry could move freely.

At the time Captain Caffrey did not know a thing about traumatic brain injuries, despite having had five or six concussions in his life from sports and other injuries. In his ignorance, he said to one of his sergeants: "So, am I screwed up because I kinda *want* to get blown up? I mean, not get hurt, just blown up, then walk away?" The sergeant said he had the same thought on his own—must be a Marine thing. Little did they know that they'd be together in the same vehicle for more than one blast. For Patrick that memory redefines the saying "Be careful what you wish for."

Before he left Afghanistan Captain Caffrey sustained three blasts, where he experienced concussions. But he thought he felt okay. After all, he reasoned, many others were much worse off than he was. However, his personality was beginning to change. He became prone to angry outbursts, a new thing for him. Arriving home, the changes became more pronounced. In Patrick's words, "I was more irritable than ever, I had intense headaches, trouble focusing and concentrating (particularly listening to what people were saying), trouble with memory, and an inability to sleep. I was rude and nasty to people and the worst part was that I didn't really know just how much I had changed."

Patrick decided to have SPECT imaging at the Amen Clinics in Newport Beach, California. "Boy, did I underestimate the value of actually looking at the brain when you have a brain problem!" Patrick said.

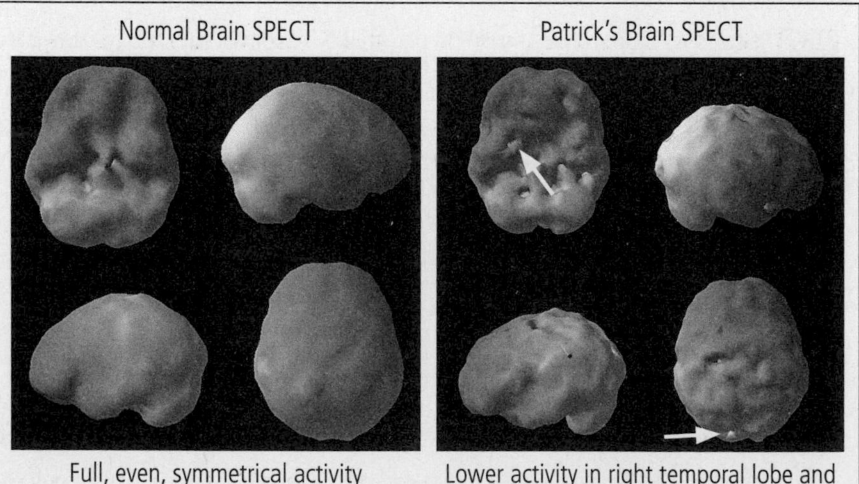

Full, even, symmetrical activity

Lower activity in right temporal lobe and left occipital lobe

We saw the damage to his right temporal lobe, which explained his behavioral and cognitive changes, headaches, decrease in ability to focus and concentrate, and memory issues.

Patrick was then put on a simple natural supplement regimen targeted to his specific brain issues. He said, "I felt a dramatic difference right away. I felt more mentally sharp and focused than ever!" Having served in the U.S. Army for ten years, first as an enlisted soldier, then as a military physician, I have a heart for soldiers; with Patrick's help, I hope to enable more servicemen and -women to get the help they need.

WHAT ALL THIS MEANS FOR YOU

So why should you care about the brains of these retired gladiators? If we can improve the brains of retired NFL players who have had tens of thousands of hits to their heads, imagine the benefit you can get with a brain healthy program, even if you have been bad to your brain. Getting on a brain-smart program can literally slow and in some cases reverse the aging process.

We have seen people improve from brain damage, brain infections, strokes, a loss of oxygen, substance abuse, and toxic exposure. The SPECT scans give us a sense of how much reserve the brain has and how much improvement is possible.

Here is a personal brain-smart program to slow aging and reverse brain damage.

1. Stop doing anything that hurts your brain. Playing tackle football at any age is not brain smart. I loved the game, but it does not love us back.

2. Focus your energies on brain-smart activities, such as those listed in this book. A healthy diet, great exercise, new learning, developing a community of healthy people, and so on.

3. Lose weight if needed.

4. Take simple supplements daily to make sure you get the nutrients you need. I recommend that all of my patients take a multivitamin and fish oil, and know and optimize their vitamin D level.

5. If damage has occurred consider these brain-enhancing supplements:

    - Ginkgo and vinpocetine to enhance blood flow
    - Acetyl-L-carnitine and huperzine A to boost the neurotransmitter acetylcholine
    - Phosphatidylserine to help nerve cell membranes
    - N-acetylcysteine (NAC) and alpha-lipoic acid as antioxidants

    I put this group of nutrients together in our nutritional supplement product Brain and Memory Power Boost, which we used with our retired NFL players. But to be clear, we used this supplement together with the whole program, which is the smartest way to use any supplementation.

6. Consider HBOT to enhance blood flow to the brain if there has been trauma. You can learn more about HBOT at www.hbot.com.

7. Consider neurofeedback to help stabilize nerve cell firing patterns in the brain. Biofeedback, in general, is a treatment technique that utilizes instruments to measure physiological responses in a person's body (such as hand temperature, sweat gland activity, breathing rates, heart rates, blood pressure, and brainwave patterns). The instruments then feed the information on these body systems back to the patient, who can then learn how to change

them. In neurofeedback, using electrodes placed on the scalp, the amounts of specific brain wave patterns are measured throughout the brain.

There are five types of brain wave patterns:

- *Delta waves* (1–4 cycles per second), which are very slow brain waves, seen mostly during sleep
- *Theta waves* (5–7 cycles per second), which are slow brain waves, seen during daydreaming and twilight states
- *Alpha waves* (8–12 cycles per second), which are brain waves seen during relaxed states
- *SMR (sensorimotor rhythm) waves* (12–15 cycles per second), which are brain waves seen during states of focused relaxation
- *Beta waves* (13–24 cycles per second), which are fast brain waves seen during concentration or mental work states

Researchers have found that people can learn to change their own brain wave states through training. Before neurofeedback is done in our office, people usually have a SPECT scan or QEEG to help guide the treatment. In our retired NFL players, we often saw excessively high slow wave (delta and theta) activity and too little fast wave (beta) activity in the front part of the brain. Many of our athletes thought of neurofeedback like going to the gym for their minds and found it very helpful.

RAY AND NANCY: A STORY OF CONTINUED HOPE

Ray White came to see us as part of our NFL study. He played linebacker for the San Diego Chargers in the early 1970s. Part of Ray's motivation for participating in our study was that his wife, Nancy, had been recently diagnosed with frontal temporal lobe dementia, and he wanted us to evaluate her as well. He was upset at the physician who diagnosed Nancy, because he told Ray that within a year she would not know who he was.

When we evaluated Ray, he showed evidence of brain trauma, as did almost all of our retired players, plus he was overweight. Nancy's scans were a disaster. She had severe decreased activity in the front part of her brain, consistent with the diagnosis of frontal temporal lobe dementia.

The feedback session, during which I showed them their scans, was very emotional for Ray and Nancy, and it was for me too. We had experience already that showed we could help Ray. But there is no known effective treatment for frontal temporal lobe dementia. My bias with cases like Nancy's is to do everything we can to try to slow or reverse the dementia process. And, certainly, it does not always work. In this case, I told Ray and Nancy that it was critical to immediately get on a brain healthy program, eat right, stop drinking alcohol, take their supplements, and get exercise, and I recommended HBOT and neurofeedback for Nancy. I also told Ray he needed to start losing weight.

Ten weeks later I saw them back for their first follow-up visit.

I was an emotional wreck the day they came back. The hour before I saw Ray and Nancy I had just found out that my work with SPECT had been attacked in a psychiatric medical journal by two doctors from the University of Texas Southwestern. I was not upset that other doctors attacked my work. By now I was used to it. I was upset that the journal's editor would allow people to write lies about me, without showing the articles to me first. I felt that was unethical. I had just gotten off the phone after having a heated exchange with the editor when Dr. Willeumier handed me Nancy's follow-up chart. I was so upset that I was shaking, which has happened maybe two other times in my adult life. At the time I was hosting five physicians from Canada who were applying our work with SPECT to their patients. I had to compose myself.

I took ten deep breaths and told myself to focus. As I calmed down, I opened Nancy's chart and didn't believe what I saw. Nancy's follow-up scan showed dramatic improvement. I showed the visiting physicians, who were amazed. I pulled up the scan on the big computer monitor in my office. It was clear that in ten weeks Nancy's brain was getting better.

Nancy had followed through on all of my recommendations for eating right, taking her supplements, and eliminating alcohol, and she had forty hyperbaric oxygen sessions and sixteen neurofeedback sessions. She had significantly improved. Her memory and cognitive function were better, her personal grooming had improved, and she was doing better taking care of their home. Ray joked that we had to slow down, because soon enough she would be smarter than he was. In addition, Ray had lost 30 pounds! He said his motivation was to help his wife. If he did everything we suggested, then she would too. They would do it as a couple. Sometimes motivation is about love. Ray loved Nancy.

Nancy: Frontal Temporal Lobe Dementia (looking down from the top)

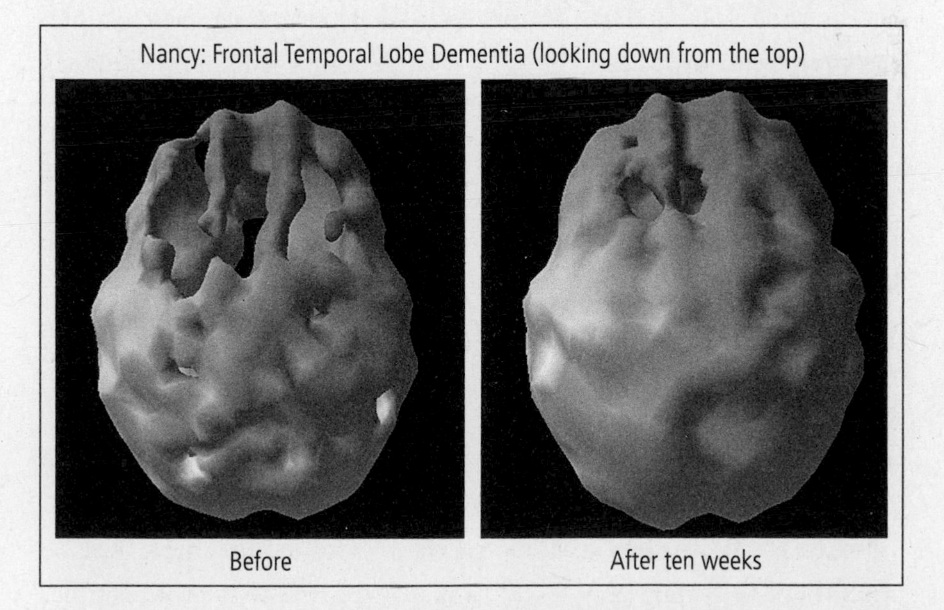

I have noticed that whenever my work is attacked there is always a case or series of cases that keeps me centered on the importance of this work. Funny how that happens.

These scans took place in early February and May of 2010. As I write this I have seen Nancy every couple of months since then; now, eighteen months later, Nancy maintains the gains she has made. She has had nearly two hundred hyperbaric oxygen therapy treatments and continues with her supplements and neurofeedback. She eats healthy, stays away from alcohol, surfs three or four times a week, goes to the gym, and recently took up singing lessons. I am in awe of this couple. I also know this program will not work for everyone diagnosed with dementia. Not everyone has a brain that will respond and not everyone can afford all of the treatments Nancy had. But I sincerely believe that we should be applying these techniques to a lot more people to recover them as much as possible. I know if it were my wife or my mother I would recommend the exact same program.

CHANGE YOUR AGE NOW: TWENTY BRAIN TIPS TO REVERSING BRAIN DAMAGE

1. Whenever brain damage is present, have a sense of urgency to repair or optimize the brain.

2. Brain aging, all by itself, lowers blood flow to the brain. As we age, our brains become more vulnerable and thus in more need of protection.

3. Use brain imaging and neuropsychological testing to understand the specific vulnerabilities and deficits in the brain. You cannot change what you do not measure.

4. Immediately stop doing anything that hurts your brain.

5. Immediately get on a brain healthy program.

6. Head injuries from your past, from taking a fall on the playground at age five to a concussion in sports to auto accidents, can alter your brain. Think back on your brain history since childhood: Was there any change in your mood or behavior following any brain trauma, even a mild one? Write them down. If so, you may want to consider some of the brain rehabilitation actions discussed.

7. When behavior, emotions, or memory is troubled, make sure to ask yourself ten times whether or not a brain injury in the past may be part of the problem.

8. Anxiety, depression, attention problems, obsessiveness, memory problems, temper problems, and low energy are just a few symptoms that could be related to a past concussion or brain injury.

9. One surprise finding during our NFL study and the brain scans from it was that lower activity in the cerebellum was common. This is an area in the lower back of the brain and is the brain's coordination center. The best activities for enhancing the cerebellum are dancing, table tennis, juggling, cursive handwriting, and calligraphy.

10. If your son or daughter is going to play contact sports, in spite of the dangers, research the program and advocate for brain-protective measures. Seventy-five percent of brain injuries happen during practice. School football programs could greatly eliminate brain injuries and concussions by limiting the amount of head to body contact during weekly practice. Thanks to research like ours and publicity about concussions, more schools are looking at brain safety in sports.

11. Many people who used to work out or played sports at a high level when they were younger have their muscles turn to fat when they give up the hard workouts. Excess weight, in and of itself, can cause lower cognitive functioning. Remember: "As your weight goes up, the size of your brain goes down." Getting on a healthy weight-loss program can improve your brain's functioning.

12. Sleep apnea often accompanies obesity. Because sleep apnea can increase the possibility of dementia, it needs to be treated. Losing weight will help, but you may also want to get a sleep study done and consult with a sleep specialist.

13. The good news is that you can reverse the brain's aging process, and even improve brain injuries, with a brain-smart program, even if the brain damage occurred decades earlier.

14. The basics of brain rehab include optimal nutrition, exercise, special coordination exercises, weight-loss classes (if needed), and supplements like fish oil and our specially formulated Brain and Memory Power Boost.

15. Many of my patients have opted to undergo HBOT, which we have seen improve blood flow to the brain. Over the years I have been impressed with HBOT's ability to increase blood flow to damaged brains. If you've had past brain damage, you may want to consider some sessions in an HBOT chamber, available in many cities now.

16. We also used neurofeedback to help reverse brain damage with great success; this involves using electrodes to measure brain electrical activity. With practice and coaching we can teach our clients how to change their brain's activity and alter brain waves for the better.

17. Brain Gym exercises can help in rehabilitating the brain.

18. Brain-enhancing supplements may be helpful for those working on repairing and improving their cognitive skills. Some of these are ginkgo and vinpocetine to enhance blood flow; acetyl-L-carnitine and huperzine A to boost the neurotransmitter acetylcholine; phosphatidylserine to help nerve cell membranes; and NAC and alpha-lipoic acid as antioxidants. (I put this group of supplements together in our Brain and Memory Power Boost.)

19. Even if you have been bad to your brain, we have seen that many, many people can recover brain function with a smart program.

20. The best time to start reversing brain damage is *now*!

9

THE TALE OF TWO RICKS

CREATE YOUR OWN GENIUS NETWORK TO GET BETTER TOGETHER

When health-conscious friends improve their health, their friends' health improves as well.

Rick Cortez is the graphic artist who makes my public television shows look so beautiful. He and I have known each other for years. Rick is thirty-one years old, a very talented, hardworking, sweet man. Since I have known him over the last seven or eight years, his weight just went up and up until he ballooned to 350 pounds. I encouraged him to get healthy, but not much happened. Then, a few weeks after we shot my last public television special, he wrote me the letter below.

Dr. Amen,

I wanted to let you know of an exciting change in my life since the taping of your new public television special. In the 5 or 6 weeks since the live recording, I've lost 30 pounds and counting.

At the time of the taping I had over 350 pounds weighing down my frame. Fast food was a staple of my diet, and portions were always big. Despite the large amount of food I'd take in daily, I had regular cravings for more of the same. I loved the "high" that came with double cheeseburgers for dinner and ice cream for dessert.

After your program I honestly didn't expect a life change. I knew myself too well—impulsive, no willpower, and no endurance . . .

But in the days following your program's taping, Marco, a co-worker of mine, decided to follow one of the Solutions you offered: "Influence others to be thinner, smarter, and happier . . ." He asked me if he could join with me to help create a healthy lifestyle, in line with the principles you laid out.

Marco then showed me before-and-after photos of a friend of his who had lost more weight than I was hoping to lose (150 pounds). The key to his sustained weight loss was a lifestyle change, not a fad diet.

That was all I needed. Suddenly it was no longer impossible, it was inevitable—*I knew I was going to lose this weight.*

About a year from now I'll be at my ideal size. But I'm in no rush—I'm having a great time getting there!

Thanks for inspiring my co-worker. In my case, it made all the difference.

Rick

The last time I saw Rick he had lost 97 pounds!

According to Rick, Marco was the kind of guy who enjoys seeing others succeed, so about a week after the final taping of my special, he told Rick, in passing, "Hey, both you and I could stand to lose some weight. What do you think about trying this Amen Solution together? You want to give it a shot?" Rick had nothing to lose, except 150-plus pounds, so he agreed. "It was having a friend who I would check in with every day on my progress, just two or three minutes sometimes, that made *all* the difference in keeping me motivated."

He's more than halfway to his goal weight and not only does he look ten years younger, there's a bounce in his step and an obvious feeling of self-confidence that radiates from him. What is so inspiring to me is the domino effect of influence that has happened simply because of one friend encouraging another. One person lost 150 pounds and inspired Rick's friend, Marco. Marco reached out to Rick, and now they are both reaping benefits of mutual support. Now Rick is inspiring his family, co-workers, and all who are reading his story here. We don't change in a vacuum. We need each other.

In a recent visit, Rick shared a little more about his journey. "I remember the weight began to creep on after I left home (and Mom's healthy cooking) for college and discovered fast food, late nights, and a sedentary lifestyle. In high school, I was active in sports, but my chosen degree and career path required many hours in front of a computer screen. I'm a huge film fanatic too, so instead of walking or working out after a day of work, I'd grab a double cheeseburger for dinner, then sit down to watch a movie with a pint of Ben & Jerry's. Did you know there's a thousand calories in one of those babies? I gained all this weight in 10-to-20-pound increments, one year at a time."

Rick went on to share the loss of energy and the desire and ability to move; he began to take walks and dance with his lovely wife of four years (who *loves* to dance!). He hired a personal trainer during one serious attempt to lose weight but found himself even more hungry, and the effort expended to the amount of pounds lost hardly seemed worth it. He settled into a state of stable misery in terms of his health. "I'm generally a happy guy, but in this area I'd simply accepted defeat," Rick admitted.

WHY WE NEED EACH OTHER

Why would you want to recruit others in joining your efforts to look and feel younger? Because they will help sustain you in the inevitable vulnerable times. There is victory in numbers and regular support. You can find friends, family members, co-workers to gather with, or join our online community at www.theamensolution.com.

I can't emphasize this enough: Social support is one of the main key elements of success! Many studies have shown that positive relationships strengthen health and longevity, while a lack of social connectivity is associated with depression, cognitive decline, and earlier death. *In one study of more than three hundred thousand people, researchers found that lack of strong relationships increased the risk of premature death from all causes by 50 percent! The health risk of being socially disconnected compares with smoking fifteen cigarettes a day and is a greater threat to your longevity than being obese or not exercising.*

What makes positive social connections so effective? Researchers have found being part of a connected community helps relieve chronic stress, which contributes to obesity, memory problems, heart disease, gut problems, insulin dysregulation, and a weakness in the immune system. When you are in a group of people with mutual care and trust, stress-reducing hormones are elevated. Women, who so naturally get together with one another to talk, chat, and bond often feel a euphoric kind of peace after a time together. Studies have shown that oxytocin, the trust hormone, is released when women share and bond. Healthy love between people is medicine that helps you live longer. Another benefit of healthy social relationships is that when your thoughts get negative or become unreasonable, healthy friends and family will give you realistic, positive feedback. Without proper feedback from others, we're all more susceptible to believing negative thoughts, which contributes to depression and diminished health.

THE KIND OF COMPANY YOU KEEP

Who you spend time with also matters. You've got to be selective because people affect your brain, mood, and physical health. A number of studies report that if you spend time with people who are unhealthy, their habits tend to be contagious. A study published in the *New England Journal of Medicine* found that one of the strongest associations in the spread of illness is who you spend time with. The study was conducted using information gathered from over twelve thousand people who had participated in a multigenerational heart study collected from 1971 to 2003. The study showed that if a person had a friend who became obese, that person had a 57 percent higher chance of becoming obese themselves. This number went up to 171 percent if both friends identified the other as strong friends. Friendship was apparently the strongest correlation: It didn't matter how far away the friends lived, as geographic distance proved a negligible factor. Sibling influence was also ranked high, with an increased 40 percent chance of becoming obese if another sibling was obese.

The study highlights the social network effect on health issues and makes an important point: Our health is heavily influenced by many factors, not the least of which is the role models around us. The powerful influence of friendship works both ways. *Researchers also found that when health-conscious people improve their health, their friends' health improves as well.* By taking the information in this book seriously, you can influence your whole network of friends and family. If you lead the way to better health in your circle of friends, your friends may also benefit. The author of the study said, "People are connected, and so their health is connected." People can connect to improve their lives through walking groups, healthy-cooking groups, meditation groups, new learning groups, and so on. When you spend time with people who are focused on their health you are much more likely to do the same.

Engaging others to be healthy is a win–win. It's helping you, and it is helping them. Just as when we get emotionally healthy as individuals, our relationships improve; so it is with physical health. If we get physically healthy, it tends to be contagious and our relationships improve in terms of more activity, eating better, feeling well, looking younger. What a gift to receive, give, and share!

> A large Swedish study of people ages seventy-five and over concluded that dementia risk was lowest in those with a variety of satisfying contacts with friends and relatives.

IS YOUR CHURCH, BUSINESS, SCHOOL, HOSPITAL, OR FAMILY A FRIEND OR AN ACCOMPLICE?

Did you know that the Cleveland Clinic, a hospital known for its innovative technology in the field of medicine, has one of the world's busiest McDonald's per square foot on its premises? Does this strike you as an obvious conflict of interest? As I started writing this book I went to an appointment with my wife to see her endocrinologist. He had bowls of candy and cookies in the waiting room. So let me get this right. Sick people go to the doctor, or a well-known medical clinic, and they're invited to snack freely on food that makes them sicker. Unbelievable! Over the last decade as my work has focused more on the connection between physical and emotional health, I have realized that many schools, businesses, hospitals, and churches could do a much better job of helping people they serve.

In August of 2010 I went to a church near my home with my family and told my wife I would save us seats while she took our daughter to children's church. As I walked toward the sanctuary, here's what I passed:

Doughnuts for sale for charity
Bacon and sausage cooking on the grill
Hot dogs being prepared for after church

As I found a seat, the minister was talking about the ice-cream festival the church had the night before.

I was so frustrated that when my wife found me in church I was typing on my phone, which she absolutely hates, and she gave me that look that only your wife can give you: *Why are you on that thing in church?* Then I showed her what I was writing:

Go to church . . . get doughnuts . . . bacon . . . sausage . . .
hot dogs . . . ice cream. They have no idea they are
sending people to heaven early!

Nearly everywhere we turn there is evidence that multiple institutions of our society, including our public schools, churches, and doctors' waiting rooms, however well meaning, are hurting us with the food they offer. There has to be a better way. Churches, businesses, schools, hospitals, and all of our other social institutions have the potential to be powerful positive influences on our health and connect us with the kind of support network that leads to success. We have to do more to make that happen.

For me, church was the obvious place to start. During that service I prayed that God would use me to help change places of worship. The house of God, no matter what religion, should not be a place that fosters illness.

Two weeks later Pastor Steve Komanapalli from Saddleback Church called me. Saddleback is one of the largest churches in America with about thirty thousand members and ten campuses across Southern California. Pastor Steve is Rick Warren's personal assistant. Pastor Warren is the senior pastor at Saddleback and author of *The Purpose Driven Life,* which has now sold over thirty-five million copies worldwide. During the 2008 presidential election, Pastor Warren and Saddleback Church hosted a civil forum with Senators John McCain and Barack Obama. Pastor Warren also gave the invocation at the 2009 inaugural address and he was on the cover of *Time* magazine with the caption "America's most powerful religious leader takes on the world." His positive influence has crossed denominational and political boundaries.

Steve asked if I would talk to Pastor Warren about a new initiative at Saddleback Church called Decade of Destiny. The staff was putting together a ten-year plan to get the church healthy physically, emotionally, cognitively, financially, vocationally, and relationally. Would I be willing to help with the initiative to help the people at Saddleback have a better brain and a better body?

I was a little stunned with how quickly my prayer of two weeks earlier was being answered. Steve set up a time for me to talk with Rick.

I found Pastor Warren to be warm and friendly. He laughed easily. But he had a serious goal: to help his parishioners (including himself) get healthier on every level. If it worked at Saddleback, he hoped he could export the plan to churches around the world (Saddleback is connected to four hundred thousand churches around the globe), as he had done with previous initiatives. To increase the health of his congregation, Pastor Warren put together a team of experts. He had already recruited noted physicians and bestselling authors Mehmet Oz (heart surgeon) and Mark Hyman (functional medicine specialist). He hoped I would provide guidance on brain health. I told him, "Count me in! Your call is an answer to my prayers about the need to change churches." I have been a Christian since a small child. I grew up Roman Catholic, was an altar boy, served at Mass when I was in the U.S. Army as a young soldier, and went to a Christian college and a Christian medical school. The project felt like home for me.

During our conversation, Pastor Warren asked, "Is there anything I can do for you to thank you for helping us?" I was just getting ready to shoot my public television special *The Amen Solution—Thinner, Smarter, Happier with Dr. Daniel Amen* and asked if he might be able to gather me an audience for a practice run. "No problem," he said, and we set a date for the following week. Pastor Warren asked if he could interview me after I completed the rehearsal and play it at the kick-off of the health portion of the Decade of Destiny program. I readily agreed.

On the day of the taping, I got to meet Pastor Steve in person outside the media center. Of East Indian heritage, his skin was the warm color of a strong latte. His dark eyes were kind and his laugh easy. I liked him immediately. He was 5'8" and about 300 pounds, however, and I hoped my work would help him get healthy.

The food in the green room was awful. There were candy bars, sodas, muffins, and pastries. I asked Steve if they were trying to kill the pastors with the food they were serving. He laughed and said, "If you think this is bad, I run a Saturday morning men's Bible study group and give the guys barbecued ribs as a reward for learning Bible verses." I was beginning to understand why God answered my prayer. With the current mentality of enticing or rewarding its parishioners with junk food, it was like one giant coronary just waiting to happen. I could also see that changing this mentality was *not* going to be easy.

The auditorium was a great place to practice my new show, and the audience loved the program. Afterward I met Pastor Warren, a very large man, both in stature and weight. I was in the middle of our NFL study so I was used to standing next to people who were 6'4" and 300 pounds, but Rick did not look healthy or vibrant. He looked tired and sick.

When we sat down for the interview, Rick started by asking me three questions in quick succession. One of the questions was about my work with ADHD (his rapid-fire questions now made sense). We talked about stress and about how increased exposure to the stress hormone cortisol puts fat on your belly and kills cells in the major memory centers of the brain.

He then asked me about the dinosaur syndrome, which I had talked about in the new show. I'd shown a slide that said:

Dinosaur Syndrome
Big Body. Little Brain. Become Extinct.

"That really got my attention," Rick said. "Can you explain that some more?"

"Sure," I replied. " 'Dinosaur syndrome' is a term I coined after reading Dr. Cyrus Raji's research from the University of Pittsburgh, which reported as your weight goes up, the actual physical size of the brain goes down. The researchers found that when subjects had a BMI between 25 and 30, considered overweight, they had 4 percent less brain volume and their brains looked eight years older than healthy people's. When subjects were obese with a BMI over 30, they had 8 percent less brain volume and their brains looked sixteen years older than healthy people's. In a follow-up study from my research group at the Amen Clinics published in the journal *Nature Obesity* we found that as a person's weight went up, the function in the prefrontal cortex [PFC], the most human, thoughtful part of the brain, went down."

"Is that why my sermons are getting longer?" Rick joked. The audience chuckled, and then we moved on to the topic of motivation.

"What moves you?" I asked Rick. "Why are you doing this new initiative?"

His answer was precise. "I want the next ten years to be the best ten years both for myself and for the church to get healthy."

We then talked about his diet. He volunteered, "I am not hungry until two in the afternoon. I could fast until noon every day of the week, but then my appetite kicks in and I eat large quantities of food until late in the night."

"You have to stop that eating pattern," I said. "Study after study has shown that people who eat breakfast are more likely to lose weight and keep it off. By eating regularly you keep your blood sugar more stable throughout the day. Stable blood sugar wards off cravings. Keeping blood sugar stable doesn't just help weight loss; it also helps your focus, memory, and decision-making skills."

The interview was fun and pleasant to that point. But then it seemed to take a strange turn. Rick asked me to give the audience some tips about brain health.

I said, "It's not magic; it's simple mathematics. If you want to be healthy you cannot eat too many calories and the calories you choose need to be of high quality. Otherwise your body and brain become bankrupt. I sent you an e-mail a while back saying if you really wanted to get the church healthy you could start by putting the calories and nutritional content on the food you serve at Saddleback. When I didn't hear back from you, I figured you weren't too keen on that idea."

This is where Rick seemed to become irritated with me. "I read that e-mail and thought, 'Oh, yeah, that's a great idea . . . I am going to become the health nut and the Gestapo for food at Saddleback.'"

I replied, "This would be one of the most loving things you can do for your church. But you have to buy into the concept, on a real emotional level, that if you overeat you are not being a good steward of your body. I can see we need to do a little therapy around this topic."

"But we built this church on doughnuts!" he replied.

Now I was horrified. This helped to explain new research from Northwestern University that reported people who frequently attend religious services are significantly more likely to become obese by the time they reach middle age. The traditions of potlucks, ice-cream socials, pancake breakfasts, spaghetti dinners, and doughnuts to get people to stay at church longer are clearly not good for the brain, body, or soul. When your brain is sick, your soul is not at its best. We've got to get creative with alternative social activities and healthier food in our churches.

I left the interview feeling unsettled. Rick was asking for help but seemed resistant to it at the same time, the same way many addicts I have treated react when confronted with the truth. "It is a process," I told myself. "Be patient."

THE DANIEL PLAN TO CHANGE THE HEALTH OF THE WORLD THROUGH CHURCHES

Over the next three months the staff and the other doctors and I developed the Daniel Plan, named after the prophet in the Old Testament who refused to eat the king's bad food.

In the first chapter of the book of Daniel (1:3–16), Daniel and his three enslaved friends, Shadrach, Meshach, and Abednego, along with other young men, were commanded to eat from the king's kitchen of rich foods and wine. Daniel and his friends were determined not to defile themselves by eating the rich food and drinking the wine. Daniel asked the chief of staff, Melzar, for permission not to consume these unacceptable foods. But Melzar implored Daniel to do as he was told, so that he, Melzar, would not be beheaded for going against the king's orders to feed the malnourished-looking Daniel and his friends.

Daniel then gave Melzar this challenge: "Please test us for ten days on a diet of vegetables and water. At the end of ten days, see how we look compared to the other young men who are eating from the king's food. Then make your decision in the light of what you see."

Melzar agreed to Daniel's challenge and tested them for ten days. At the end of ten days, Daniel and his three friends looked healthier and better nourished than the young men who were eating the food assigned by the king. So after that Melzar fed them only vegetables instead of the food and wine provided for the others. God gave these four men an unusual aptitude for understanding every aspect of literature and wisdom. Daniel and his friends looked better and were smarter than the others.

The Daniel Plan is a fifty-two-week small-group program to get the church healthy. Small groups are the secret sauce of Saddleback Church, where members meet weekly for an hour or two at someone's home or at a restaurant to study a specific topic, such as a book in the Bible. It's the secret sauce because social and community support is the key ingredient to any real change. You cannot do it alone. These small groups enhance commitment and learning and provide ongoing encouragement and emotional support. Saddleback has about five thousand small groups, and the plan was to use this format to maximize results and help the church get healthier. The prophet Daniel had his posse of like-minded supporters and you should too.

> Research shows that those who have the highest levels of social activity experience one quarter the amount of mental decline in their golden years as those who are not at all socially active.

In November and December Pastor Warren talked to his church extensively about the Daniel Plan. On December 12, with my wife, Tana, and me in the service, Rick told the congregation that he was coming after them to get healthy on January 1. I was so pleased to see the progress of the pastor and the church. But then he said something that completely baffled me.

"But between now and January first, eat anything you want!"

I looked at my wife Tana in disbelief. "Did he really say that?" I asked. If I had a tomato, I might have thrown it at him. For real change to take place, the new behavior cannot occur sometime down the road. It needs to start now, not some date in the future.

Shortly after the service, Rick and I had a chat about the comment. Almost immediately he got the idea. "So you're saying what I did was kind of like telling a young man that since he is about to get married, he should have a few last flings before he ties the knot."

"You got it," I said. "If people are serious about wanting to change, *now* is the best time to start, not tomorrow, Monday, or January 1."

For Christmas I bought Rick a first edition of C. S. Lewis's parable *The Great Divorce*, a wonderful book about lasting change. I highlighted the following passage: "The gradual process is of no use at all . . . This moment contains all moments." For change to occur, you need to have a sense of urgency, *in this moment right now.* That is why many people decide to get healthy after having a heart attack or being diagnosed with cancer. My hope for you is that you do not need a crisis to start to get healthy and that the value of avoiding a crisis has enough emotional motivation.

The formal start of the Daniel Plan kicked off January 15, 2011, with a big rally at Saddleback Church. It was an enormously popular event and the church had to turn away thousands of people. The excitement was palpable. We'd designed a brain-smart curriculum and ninety-two hundred people signed up to be in our research study. On this day, Rick weighed in at 292 pounds. By the time of our third rally in October, Rick had lost 50 pounds and ten inches off his waist, looked healthier, and . . . ten years younger! He told the congregation that his secrets included:

Rick

Before

Down 50 pounds after ten months of the Daniel Plan

1. Focusing on his motivation every day and deciding to view physical and emotional health as a spiritual discipline. He often repeated this New Testament verse: "Do you not know that your body is a temple of the Holy Spirit, who is in you, whom you have received from God? You are not your own; you were bought at a price. Therefore honor God with your body" (1 Corinthians 6:19–20).

2. Keeping a food journal so he knew what he was putting in his mouth. He said it was quite a wake-up call.

3. Drinking water throughout the day.

4. Getting proper sleep. Until we met, Rick spent a day or two a week up all night with no sleep at all. He never knew it was a problem, but research suggests that when people get fewer than six hours of sleep at night, they have lower overall blood flow to the brain, which means more cravings and more bad decisions. Other research suggests that sleep deprivation causes people to be unrealistically optimistic and engage in riskier behaviors. To look and feel younger it is critical to focus on getting enough sleep, which usually means more than seven hours a night.

5. Consuming high-quality calories. Rick dumped the junk food and focused on only eating high-quality food. He told the congregation he eliminated the four white powders: cocaine (he was kidding), sugar, bleached flour, and salt. He eliminated doughnuts from his diet and started each day with a healthy breakfast. He ate smaller meals throughout the day. He stopped drinking his calories (especially sodas) and using artificial sweeteners, and he dumped bread from his diet, as it immediately turns to sugar in the body. Eating this way curbed his cravings. He also started eating slower to enjoy his food more and increase his sense of feeling full faster.

6. Getting regular exercise, which included weight lifting and cardiovascular work. He was being much more consistent with his trainer.

7. Taking simple supplements, such as a multivitamin, fish oil, and vitamin D.

8. Being accountable to his weekly small group.

All of the components were necessary for the plan to work, and they are all necessary for you to look and feel younger, but it was the small-group component that turned out to be the secret sauce that really made the whole thing work. *When you do this program with another person, your family, a group of people at church, work, or within your community, the healing process becomes much more powerful.*

I've no doubt, with a team of support and the program outlined in this book, you can do this too.

Whenever I am on the Saddleback campus, I hear story after story of how people's lives have been changed, using the principles in this book in a small-group format, where people gather together weekly to support each other. People have told me:

"I've lost 20 (or 30 or 60 or 90 or 150) pounds."

"My numbers are so much better!"

"No more headaches! It's amazing. I was taking prescription pain medication almost daily, and now it's been more than two weeks without any pain or pills!"

"My clothes fit loose and I can get back into my old ones."

"Color is coming back to my gray hair . . . who knew?"

"My mood is so much more stable and positive."

"My asthma is better."

"With the elimination of sugar, flour, salt, and processed foods, I rarely have any cravings and have found I eat smaller amounts of nutrient-rich foods."

"My husband also lost 25 pounds!"

"I just finished chemo. Everyone is amazed at how much energy I have and how fast my hair is returning. I am running circles around a friend who is ten years younger who doesn't have cancer. (He is not on the plan.)"

"My complexion looks great; the improvement in the smoothness of my skin is remarkable."

"I've been off wheat for six weeks, and no more acid reflux."

"Ninety-eight percent of my headaches at night have disappeared. I wake up feeling clear-headed instead of foggy."

"I don't have body, joint, muscle pain in the mornings."

"I'm off my high blood pressure meds . . . and am working on getting off my type 2 diabetes and cholesterol meds."

"I am diabetic. Now my blood sugar is dramatically better than when I was on insulin. I am not taking either now."

"I'm having less arthritis (inflammatory) pain."

"I lost 3 inches off my waist and 4 inches off my hips."

"I have smoother, healthier facial skin with reduced acne."

"I have fewer PMS symptoms."

"I enjoy the adventure of discovering new foods, cooking new meals, and trying new things at restaurants."

"My triglyceride levels have lowered; I have reduced joint pain by at least 90 percent and am no longer taking that harmful medication! Oh, and I can play with my grandkids!"

"Odd to say this in church, but my sex life has dramatically improved!"

THE SECRET SAUCE IN PRACTICE

It is Sunday afternoon, and Cindy, a single working mother of four young kids, and a recent cancer survivor, is in her kitchen surrounded by several other women friends. All have full-time jobs, and all decided to get healthy together. The house is filled with delicious smells and the sounds of happy kids in the background. It's clear these women are having a blast cooking healthy meals together for the whole week ahead. One friend is making brain healthy nutritious mini breakfast casseroles in muffin tins that will be warmed up on busy school and work mornings to come. The jobs—grocery shopping, menu planning, cooking, and storing the meals—are divided up between the women according to their preferences. What was once an enormous chore for these single moms and working women is now fun, and the time saved by pooling their efforts on meals gives them much needed downtime to rest and be with their kids.

During her cancer treatments, Cindy was given steroid drugs and gained 30 or 40 pounds, which her body hadn't been able to shed. In fact, all the drugs she took for her cancer therapy had still not detoxed completely from her body. Because Cindy and her friends were no longer on a diet, but on a brain healthy lifestyle, she is now consuming foods that are actually helping her body heal itself. Her weight is falling off, her energy is returning, and she's feeling the old Cindy coming back to life. But Cindy would say it is the camaraderie that makes all the difference—not only in helping launch a new way of eating but in sustaining the changes. In addition, the women meet to walk and talk every week, encouraging each other spiritually and emotionally.

Dee Eastman, a friend of the Amen Clinic for over a decade and the director of the Daniel Plan at Saddleback, enthusiastically shared Cindy's story above, along with many others like it. As you know by now, support in the form of teams or groups is a major key to your success if you plan to look and feel younger. The research is clear that people rarely change in isolation; it tends to happen most often in small groups.

Dee said she expected to see people's weight drop and their bodies get more fit, but remarks, "I was literally blown away by the amount of extra health benefits that were pouring in and being reported to us in such a short time. Within just three months, the folks who immediately jumped in, got their blood work done, and ran with the program began seeing radically better medical lab results. Many were able to get off blood pressure and cholesterol medication. Others reported being able to sleep through the night without the need for sleep aids." She is seeing with her own eyes what I've been saying for years: "Food is medicine." (For more on this, revisit chapter 2.) In addition to better lab numbers, Dee is hearing reports of depression lessening, anxiety fading, moodiness calmed. Dee admits that such impressive changes, across a wide spectrum, would not have occurred without the secret sauce of mutual support via small groups and online support groups.

Here are a few observations of the positive dynamics for change that Dee has observed happening because of friends supporting friends in getting healthy and functionally younger.

Health is creatively incorporated into visiting and social gatherings. One active, intergenerational group (one of the couples is seventy-nine years young) meets to walk and talk together three times a week, then meets on Saturdays at a local gym to do circuit training. As they walk or work out, they casually share healthy recipes and cooking tips, along with other life joys and struggles. They are finding it is possible to get connected and get healthy at the same time.

Inner and outer healing go hand in hand, particularly in a supportive group culture. There is a woman who had almost thrown in the towel on life, who found refuge and encouragement in a group. She was about 5'2" and weighed in the mid-300-pound range. She dove into getting healthier with her whole heart, feeling buoyed by the nutritional and exercise plan and supported by her new friends. Not only has she lost 40 pounds (the first time she's been under 300 pounds in years), but she's also healing emotionally while getting better physically. (It makes one wonder about possibilities for therapists doing counseling while they walk instead of sit with their clients!) The exercise has helped her depression with good endorphins and blood flow and body composition, and in Dee's words, "being enfolded into a loving community" is healing her heart and soul. Human beings were created and wired to grow best in community, emotionally, spiritually, and physically.

Healthy support groups create a safe environment for being honest with ourselves and each other. Dee credits Rick Warren with leading the way in being truthfully vulnerable. "It was really freeing when Pastor Rick admitted, 'Hey, I got off-track in my health. I've been focusing on trying to save the world, but I've neglected my physical body. Now I have ninety pounds to lose.' Rick also reminds his congregants often that they don't have to go through anything alone. He's in a small group that is encouraging and holding him accountable alongside the rest of the church.

Groups can create change in the kind of eating that happens in social situations. One group meets at a well-known healthy restaurant once a week to enjoy a nutritious meal together. People are bringing brain-smart foods as refreshments to meetings and events. One group is planning a healthy potluck, where each member is going to bring their favorite brain-smart dish, along with the recipe and nutrition information.

> Studies show that overweight and obese young adults who had more social contacts trying to lose weight were more likely to want to lose weight themselves. Encouragement and approval from social contacts account for this association, researchers say.

SMALL-GROUP TIPS FOR SUCCESS

Debbie Eaton is an expert in small groups. She oversees seven hundred of Saddleback's small groups and has done so for many years. She offered some tips for making groups succeed. You may want to consider these as you join or create your own support group to begin implementing this program into your life.

HIGH ACCOUNTABILITY AND HIGH ENCOURAGEMENT

Problems in groups almost always occur when one of these traits is out of balance. Too much accountability without plenty of cheerleading leads to feeling discouraged or feeling as though group members are pushing toward perfectionism. Lots of encouragement, without the balance of holding each other's feet to the fire, leads to groups that stagnate and do not stretch or grow.

SHARED PASSION

It is important that those who join your group share a passion, such as passion for physical well-being. Small groups often work best when two good friends invite other people they know, so there's some sort of connection already in place. Perhaps you have one good friend who is "all in" with you in this journey to health. If both of you can think of one or two other friends who are passionate about losing weight, feeling good, and growing younger, the group can form and expand quickly and organically.

SIZE MATTERS

The ideal size for a support group is eight to ten people. If the group gets any bigger than this, the introverts may shut down and extroverts take over. Also, after a group exceeds ten people, they start thinking the group is so big now that their occasional absence won't be missed. People aren't as consistent and accountability is not as high. On the other hand, if a group is too small, over time it can stagnate and seem to get off kilter and ingrown. There aren't enough personalities or a wide enough experience base to keep the dynamics lively and vibrant.

If your purposes are smaller and more specific—to exercise regularly or check in with a nutrition coach daily, for example—it may be served by just one or two committed friends, such as was the case with Rick and Marco.

BEGIN WITH LIMITS IN MIND

Generally, it is best to commit to a group for a set period of time. You may try starting with a six-week commitment, and then re-up if the group is working well for everyone.

In addition to these ideas, here are a couple more researched-based thoughts on successful support teams.

PROXIMITY HELPS

Willow Creek Church, another of the largest churches in America, realized its small groups were not working to help people feel and stay connected in their real lives. It was not until fairly recently that Willow Creek began tapping into neighborhood communities (called Table Groups), which met together several times a week, that people once again began to care, share, struggle, grow, and bond. So consider convenience and location to make meeting together happen more easily. In Rick Cortez's case, having an accountability partner at work made it convenient to check in with each other before the work day or on their lunch breaks. Neighbors make wonderful walking or jogging partners. Make getting together as simple as you can and you're more likely to do it.

BE WARRIORS FOR EACH OTHER

You have to be ruthless for your health and a warrior for the health of those you love. Research shows that when group members go soft on one another, to the point where they accept or empathize with backsliding too much, the group dynamics change. Without meaning to, the group has now become supportive of slacking off . . . and a deterioration of health occurs. Patients sometimes ask me why I am so direct and do not just simply accept excuses. This is not my personal nature, believe it or not. By nature, I'd love to be a middle-of-the-road, everything-in-moderation, just-try-your-best kind of guy. But the truth is: This is not helpful, not loving, and not in your best interest. Have you ever had a friend who cared enough not to bullshit you? That's me. If I'd let Pastor Rick continue telling people to eat anything they want, we would have started the plan tomorrow, and because tomorrow never comes, his congregation would remain sick and headed toward an early grave. Let's create a movement together. It can start with just you and a few friends.

The following story is one about using the secret sauce of unwavering support within marriages and families, so you can leave a legacy of joy and health to your children and grandchildren.

PASTOR STEVE: LEAVING A BETTER LEGACY

Remember Pastor Steve Komanapalli? As part of his evaluation at our clinic, he had a SPECT scan. His brain was not healthy and showed very low activity in his PFC, the part of the brain involved in judgment and impulse control. On cognitive testing, he also scored poorly, especially in the area of attention. He was on multiple medications for diabetes, hypertension, and cholesterol problems. At his first appointment he had to roll off the couch owing to his weight. When I first met Steve, he and his wife, Nicole, were expecting their first child, Karis, a baby girl. I liked Steve a lot, so I had to be straight with him. He could not have his chocolate cake (and barbecued ribs, buffalo wings, fried chicken, loaded pizzas, and supersized root beers) *and* his health and longevity too. He could not eat without limits and leave a positive legacy for the next generation. I told him that if he did not get serious about his health, then a stepfather would be raising Karis because he would be dead. Did he really want to put Nicole and Karis through his early illnesses and death?

Around the same time, he had a talk with Nicole, during which she told him, "If you die of something preventable and leave me to raise our child without you, I will mourn your loss, but I will be deeply disappointed that you didn't love us enough to make your health a priority. I am really going to be hurt that you didn't put your well-being, and our family's needs, above your appetite." These two conversations began Steve's ascent into health.

Steve became one of the leaders of the plan on the Saddleback campus. "I began to see that the single biggest impact on how I feel is what I put in my mouth," Steve said as he described one of his many "aha" moments. Within one month of making food changes, Steve's cholesterol and triglycerides had come down to the normal range. He is now playing Ping-Pong, which is one of my all-time favorite games for improving the brain, and when played well, a game of Ping-Pong will get your heart pumping but is also doable at Steve's size. When he's hungry, Steve snacks on fruit and nuts and finds he is satisfied with a small amount, unlike the never-ending cycle of craving and hunger that junk food perpetuates.

After five months on the plan, Steve lost 35 pounds and 4 inches off his waist. His health numbers improved dramatically:

- Triglycerides decreased from 385 to 63
- Cholesterol decreased from 200 to 130
- HDL (good cholesterol) increased from 22 to 46
- Blood sugar decreased from 128 to 89
- HgA1c (a marker of diabetes) decreased from 7.2 (abnormal) to 5.7 (normal)
- He is off both his blood pressure and cholesterol medications!

Steve's follow-up SPECT scan showed remarkable improvement in his PFC, and his attention scores had dramatically improved. His brain, body, and mind were significantly younger within just five months of starting the program. Steve is blessed to have a wife who loves him enough to challenge and support him. Sometimes the best secret sauce is right in your own backyard or looking at you from across the dining room table.

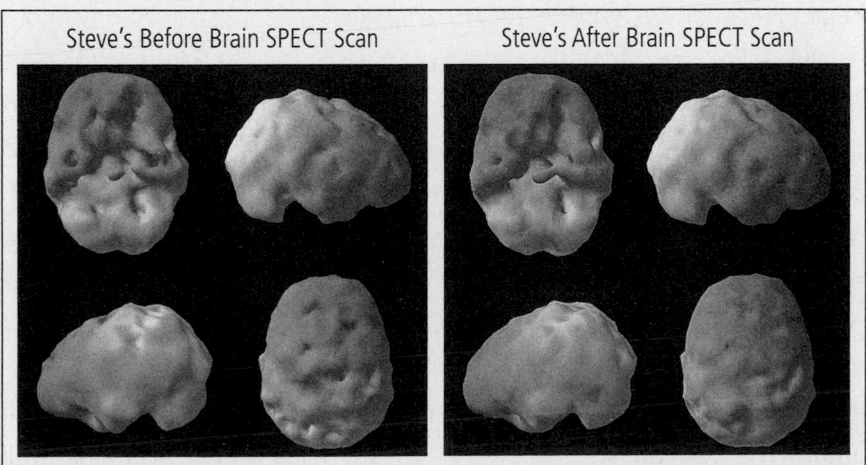

Steve's Before Brain SPECT Scan — Many areas of decreased activity

Steve's After Brain SPECT Scan — Overall improved activity

> Studies show that people in loving relationships tend to live longer, in part because they help monitor each other's health.

The Saddleback story and its combined statistics are staggeringly positive. In the first five months the church lost a total of 200,000 pounds, about 7.68 percent among our research participants. A 5 percent reduction in body weight decreases the risk of diabetes by 58 percent. Eighty percent of our participants said they were compliant or very compliant with the program. Eighty percent said it was easy or very easy to do. Fifty-five percent were doing it with someone else in their families, and 80 percent said they had increased their exercise.

However, it is the individual stories that touch your brain and inspire change. It is the real people behind the numbers that make me smile. The other day I took a walk on Balboa Island near my home in Orange County and was stopped by a number of people out for a walk who recognized me from public television. One couple, who had been incorporating these principles into their lives, had lost 60 pounds between them. Everywhere I go I meet people who express their gratitude for how this knowledge has led them to better lives.

The keys to getting healthy are absolutely important to know; but the secret sauce that makes it work and last, and adds fun and motivation to the process, is doing it together. Grab a friend or family member to do this program with you. You will both be better off.

CREATE YOUR OWN GENIUS NETWORK

A Tale of Two Ricks is about using the resources in your relationships to get and stay healthy. I am in a professional support group run by my friend Joe Polish. In a recent group he gave us an exercise on the power of networks, which he has graciously allowed me to share with you. Called "Create Your Own Genius Network," it can help you thrive and keep you on track toward your goals, in a similar way to the small groups at Saddleback. Research has demonstrated that strong relationships are associated with health, happiness, and success. The health of your peer group is one of the strongest predictors of your health and longevity. This exercise will help you create and sustain your own network.

WHAT ARE YOUR HEALTH GOALS? (BE SPECIFIC.)

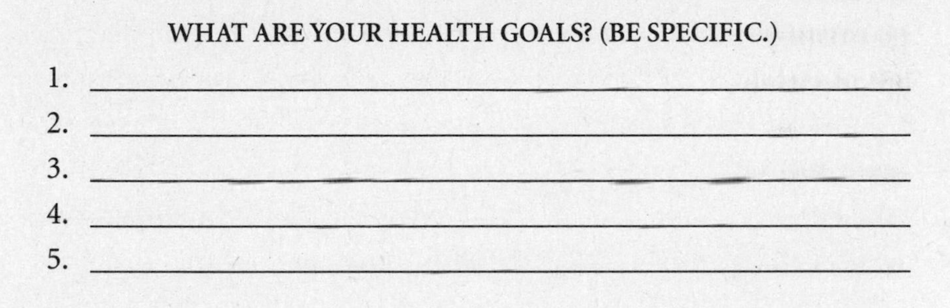

WRITE DOWN THE NAMES OF FIVE PEOPLE WHO CAN HELP YOU REACH YOUR GOALS AND BE SUPPORTIVE OF YOUR EFFORTS TO GET AND STAY HEALTHY.

WHAT WISDOM DO THEY HAVE (HEALTH ADVICE, EXERCISE BUDDY, SUPPORT, ETC.)?

HOW CAN YOU BE HELPFUL TO THEM? GIVING BACK IS A KEY INGREDIENT TO MAKING A GENIUS NETWORK WORK.

1. ______________________________
2. ______________________________
3. ______________________________
4. ______________________________
5. ______________________________

HOW CAN THEY HELP YOU? BE SPECIFIC (WALK TOGETHER ONCE PER WEEK, SHARE HEALTHY RECIPES, ETC.).

1. ______________________________
2. ______________________________
3. ______________________________
4. ______________________________
5. ______________________________

Set aside time each week to connect with the five people in your genius network, whether in person, by phone, by e-mail, or by text. If you do this one exercise, you will start to build a great network to help you look better and live a healthier, longer life. Even though this exercise is very simple, it is also powerful. Keep your genius network constantly up to date and make sure to support others in their efforts to use their brains to change their age. You will be supporting yourself in the process.

CHANGE YOUR AGE NOW: TWENTY TIPS FOR GETTING BETTER TOGETHER

1. The secret sauce for brain health and longevity is to do it *together*. Start making a list of people who will support you and vice versa. We are more powerful when we use more than one good brain at a time.

2. According to C. S. Lewis, in his short parable *The Great Divorce*, "The gradual process is of no use at all . . . This moment contains all moments." Now is the time to get well, not at some undetermined date off in the future. Choose people to join you who are ready to get healthy *now*!

3. Begin every day by focusing on your goals and planning how you'll meet them, and then share this with your accountability buddy. Short daily check-ins are powerfully motivating.

4. You have to snooze to get healthy! Sleep is a critical ingredient to longevity success. Focus on getting eight hours of sleep at night to boost brain function and follow-through. Encourage your friends to do this as well.

5. Are you a friend or accomplice? Write down the names of five people with whom you spend the most time. Are you supporting their health efforts (are you their friend)? Or, are you supporting their bad habits (are you their accomplice)?

6. Combine healthy eating with friendships. Prepare healthy meals and snacks for the week ahead with friends; share recipes and ideas for cutting calories and upsizing nutrition; bring delicious, healthy food to potlucks and parties.

7. Exercise regularly, with a partner or a group of friends. It helps if you can make this convenient by walking with people who live nearby, or working out at the gym together before or after your regular get-togethers.

8. Create a Facebook group of friends who commit to check in with what they did to incorporate exercise into their routines that day.

9. Incorporate exercise into social routines. Take an after-dinner walk with friends, meet someone to play tennis before lunch, bike to social events.

10. Create warm memories in the kitchen with your family in healthier ways. For example, instead of baking sugar cookies, let the kids decorate their own mini pizzas or fruit or veggie "art" sculptures (with bits of fruit or veggies and toothpicks).

11. Remind your kids and spouse that what they put in their mouth affects how they feel. Offer your family plenty of attractive and tasty good-mood foods that nourish their brains and bodies.

12. Determine to create a healthy legacy for your family. It starts with you showing the way. Prioritize time for active play, gardening, or shopping at farmers' markets, and cooking healthy meals together.

13. In your support group, whether it is made up of two people or ten, be sure you have a good balance of high accountability and high encouragement. Be warriors for each other's health.

14. Plan ahead when dining out for dates, social lunches, couples' nights out, and so on by scouting out healthy restaurants in your area. Or start a dinner group where you take turns hosting other couples for delicious, brain healthy meals.

15. Create a genius network using the form in this chapter, contacting five people who you think might be willing to support each other in developing new healthy habits.

16. Spend more time around healthy people in general, as you really

do become like those you spend the most time with. Healthy habits are contagious!

17. Commit to "influence others to be thinner, smarter, happier and younger." Be patient with their process but consistent with your new behaviors. Encourage every step made in a positive direction.

18. "Tell the truth in love" to someone whose health you care about. It wasn't easy for Steve's wife, who loved her husband unconditionally, to challenge him to change. But by doing so, she likely gifted them with many more healthy, happy years together.

19. Make a group goal to celebrate your success *together.* Rick Cortez is looking forward to dancing with his wife. Other groups choose to run a 5K race together, and some climb a summit to celebrate.

20. Consider the possibilities of joining an online community of support. We created the Amen Solution community website (www.theamensolution.com) to offer just such an opportunity. I'm there as your "virtual brain coach," along with others choosing the same journey toward a brain healthy lifestyle.

10

DANIEL AND BRAIN SPECT IMAGING

WHAT YOU DON'T KNOW IS STEALING YOUR BRAIN

I curse you that you know something that is true that no one else believes.

—Romanian curse

From the moment I got my own brain scanned for the first time in 1991, I developed brain envy. I had already scanned dozens of patients when I decided to get my own scan at age thirty-seven. When I saw the toxic, bumpy appearance, I knew it was not healthy. All of my life I have been someone who rarely drank alcohol, never smoked, and never used an illegal drug.

Then why did my brain look so bad?

Before I truly understood brain health, I had many of the bad brain habits discussed in this book. I played football in high school and got my bell rung on several occasions. I ate lots of fast food, lived on diet sodas, and would often get by on four or five hours of sleep at night. I worked like a nut, didn't exercise much, and carried an extra 30 pounds that stubbornly would not go away by wishful thinking.

My last scan, at age fifty-two, looked healthier and much younger than it did fifteen years earlier, which typically does not happen with the aging process. *Brains usually become less and less active with age.*

Why did my scan look better? Seeing other people's scans compared with mine, I developed "brain envy" and wanted a better brain.

Brain SPECT imaging literally changed everything in my life. It helped me come out of the darkness, both in my personal and professional life.

OUT OF MY PERSONAL DARKNESS

Never before being scanned had I thought about the physical health of my own brain, despite being my medical school's top student in neuroanatomy, completing five years of postgraduate residency training, and becoming board-certified in both general and child and adolescent psychiatry.

For example, I had no idea that:

- Being overweight negatively affected the health of my brain
- High-fat, sugary foods worked on the addiction centers of my brain
- Spraying cleaning chemicals in a closed shower stall was a really stupid idea, as it was toxic to my brain
- Getting less than six hours of sleep at night lowered the blood flow to my brain. I rarely got more than five.
- The chronic stress from working so many hours was hurting cells in the memory centers of my brain
- Being around secondhand smoke damaged the blood vessels to my brain
- Drinking a 36-ounce caffeinated diet soda, not uncommon for me pre-SPECT, dramatically constricted blood flow to my brain

In short, what I didn't know was hurting me. And not just a little. As I have learned from looking at over seventy thousand SPECT scans from

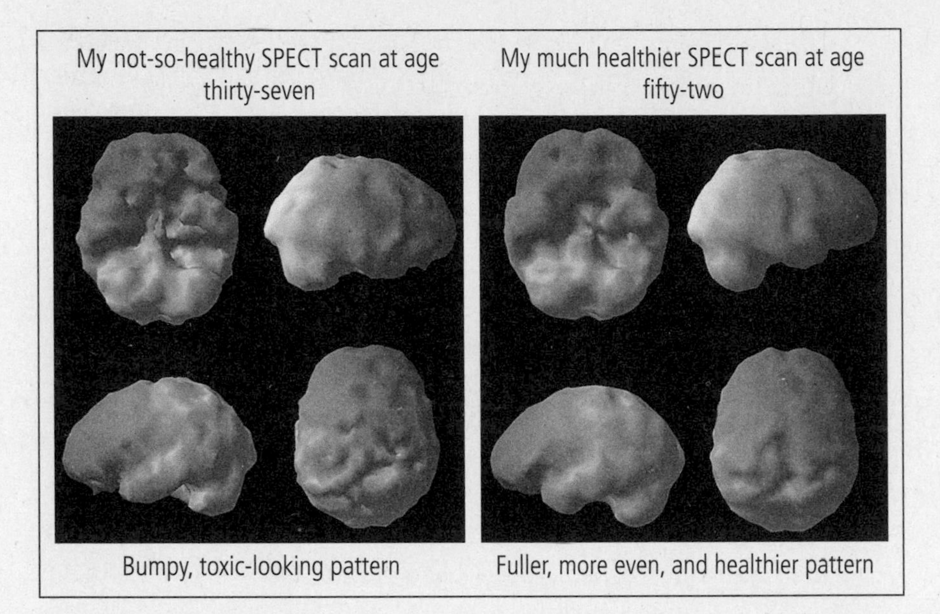

real people who were suffering, *having a less-than-healthy brain went with less-than-healthy decisions and a suboptimized life.*

Prior to my first scan I had no idea that I could actually change the health of my brain by changing my habits. I thought that some of my cognitive issues, such as dropping names or being unfocused and forgetful, were just a normal part of the aging process. After all, I was now thirty-seven years old. The dozens of decisions I made every day were being made without the knowledge of what I was actually doing to my brain.

As I learned about brain imaging and brain health, I put into practice all of the main principles in this book. I started to exercise more, improved my nutrition, got more sleep, monitored my bloodwork more often, and took targeted brain healthy supplements.

Over time, seeing the SPECT scans caused me to look at every aspect of my personal life, even the way I viewed the world, in a different way.

- Watching contact sports was definitely not as much fun. I realized that when I watched boxing or football, I was watching brain damage in process. I was witnessing athletes' brains being ruined, which in turn would ruin their lives. I remain completely baffled as to why ultimate fighting is legal. Don't the athletic commissions know that boxing dementia (dementia pugilistica) was first described in 1929, and that ultimate fighting is as bad, or worse, than boxing? These athletes get kicked or kneed in the head repeatedly.

- When I watched the news and saw natural disasters, I wondered how many survivors would develop emotional trauma patterns on their scans or would develop brain dysfunction secondary to the chronic stress.
- When I read stories about soldiers returning from war with brain injuries from IEDs I was horrified that the military was not routinely using functional brain scans to evaluate the state of their brains' function and actively trying to rehabilitate them early. I was an enlisted infantry medic and then an army psychiatrist and knew our soldiers deserved the best care possible.
- When I saw stories of people who committed terrible crimes, such as murder, rather than just judging them as bad, as is so easy to do, I wondered if they had dysfunctional brains,. I have subsequently published professional papers on our work with murderers.
- SPECT changed the habits within my family, because I wanted my wife, children, and grandchildren to have the benefit of a brain healthy life. If you dated one of my daughters for more than four months, you got scanned. I wanted to know about the health of the young men's brains.

SPECT definitely shed light on my own brain's "lack of health," and as I gained more experience with SPECT, I realized it was possible that by changing my habits I could indeed improve the overall health of my brain and subsequently change my life. I got a burning desire to make my brain better. In a sense, I fell in love with the health of my own brain. The extensive experience I was gaining with patients' scans before and after treatment convinced me even more that our habits accelerate the aging process—or they can decelerate it.

OPENING THE DOOR ON PROFESSIONAL DARKNESS

By the time I went to my first lecture on brain SPECT imaging in 1991, I had been evaluating and treating psychiatric patients for nearly a decade without the benefit of any brain imaging. I often felt in the dark on what to do for my patients, even though I was highly trained and certified competent. Whenever I saw an older person who was depressed or complained of memory problems, a treatment-resistant substance abuser, an aggressive teenager, or a couple who couldn't get along, knowing what to do for them felt like shooting craps. I would do what I was trained to do, such as give an ADHD child a stimulant medication or a depressed patient an antidepressant, and sometimes it would work, but often enough it would make the patient much worse, sometimes homicidal or suicidal.

It felt as though I were throwing darts in the dark. Sometimes I got it right. Sometimes I hurt people.

And I spent a lot of time anxious, knowing I was practicing a very soft science at best. I often wondered how well my other medical colleagues—cardiologists, orthopedists, neurosurgeons, and gastroenterologists—would be able to diagnose and treat their patients if they were unable to image the organs causing the problems.

For me, everything changed professionally after attending that first lecture on brain SPECT imaging at a local hospital in northern California where I was the director of the dual-diagnosis program, which took care of psychiatric patients who also had substance abuse problems. (*Dual diagnosis* refers to a patient with two issues, such as bipolar disorder and alcohol addiction. Mood disorders and addictions, for example, often go hand in hand.) From the moment I ordered the first SPECT for a patient, it was as if I'd been given glasses to see what was happening in the brain. No more flying blind and guessing. Of course, having a SPECT scan does not always make a dramatic difference in patient care, but my first ten cases were so helpful that I was completely hooked. Here are several examples:

- Matilda, sixty-nine, had been diagnosed with Alzheimer's disease but did not have the Alzheimer's pattern on SPECT, which was already described in the medical literature in 1991. Her scan was more consistent with depression. Her memory was restored when I treated her for depression.
- Sandy, forty-four, had a clinical profile consistent with adult ADHD (short attention span, distractibility, disorganization, procrastination, and impulse control problems), yet she refused treatment. When she saw the evidence of ADHD on her scan she started to cry and said, "You mean this is not my fault," and immediately agreed to give the medication a try. It made a dramatic difference in her life and her marriage. I already believed I knew her diagnosis, but the scan helped *her* believe it.
- Geraldine, seventy-two, was suicidal and suffered with resistant depression. On her scan she had two huge right-sided strokes in her brain that had previously been undetected. Knowing about the strokes helped us understand her depression better and also helped us prevent the possible third stroke that could have killed her.
- Chris, twelve, was hospitalized for the third time for violent outbursts. He had been seeing a psychoanalyst in the Napa Valley who wondered if the problem wasn't his relationship with his mother. His scan showed very clear trouble in his left temporal lobe, an area underneath his left temple and behind his left eye. This area is often associated with violence. When put on an antiseizure medication, his behavior normalized and he subsequently thrived in a public school setting. Without the scan, he was on his way to institutional living in a residential treatment program or jail.

- Sherrie, fifty-two, had been diagnosed with bipolar disorder but refused to take her medication. She was in her third hospitalization for hearing voices coming from the walls of her home. She had tried to pull out all the wiring in her home to get rid of the voices. Once she saw the abnormalities on her scan she became compliant with treatment and quickly improved.
- Ken, fifty-nine, was abusing alcohol and cocaine, but was basically in denial about his substance abuse. After seeing his scan he developed brain envy and completely stopped his drug abuse and adopted a brain healthy lifestyle. A year later his brain was dramatically better.
- Sara, forty-two, and Will, forty-eight, failed marital therapy multiple times. When they were both scanned, Sara had an obsessive-looking brain (too much activity in the front of her brain), while Will had an ADHD-looking brain (too little activity in the front of his brain). On appropriate treatments to individually balance their brains, their marriage dramatically improved.
- Ted, seventeen, had failed a residential treatment setting. He suffered with several impulsivity and criminal tendencies. He was missing the left front part of his brain on SPECT, which means he had zero blood flow to this part of his brain. It turned out that he fell down a flight of stairs when he was four years old, was unconscious for a half hour, and no one remembered the injury or thought it could have been involved in his troubled behavior.
- Christina, sixty-two, had been diagnosed with chronic fatigue syndrome. Her primary care doctor said she had depression and a personality disorder and sent her to me. On her scan there was a clearly toxic pattern, consistent with a brain infection. She was depressed and had a disordered personality because the organ of mood control and personality was damaged, which completely changed the treatment plan and our approach to Christina.

In a few short months SPECT completely changed how I practiced medicine. How could I practice without imaging? How did I know what was going on in my patients' brains unless I looked? My anxiety with my patients went down and my excitement about being a psychiatrist went up. My effectiveness and confidence also went up. I was more willing to take on complex and treatment-resistant cases.

- In order to effectively treat patients, I knew it was essential to look at their brains before I went about trying to change them. *You cannot change what you do not measure.*
- SPECT helped me make more complete diagnoses for my patients and not miss important findings, such as a past head injury, infection, or toxic exposure.
- SPECT helped me be more targeted in my treatments. I have learned through the imaging work that illnesses like ADHD, anxiety, depression, addiction, or obesity are not single or simple problems in the brain; instead, they all have multiple types and the treatment needs to be targeted to the specific type of brain, not a general diagnosis like depression.
- SPECT made me considerably more cautious in prescribing certain medications or using multiple medications, because they often appeared toxic on scans. I had to be more responsible in my use of medications.
- SPECT directed me to use more natural treatments, as they are often effective and appear less toxic on scans.
- SPECT helped to break denial in substance abusers. It is hard to say you do not have a problem when faced with a scan that looks toxic.
- SPECT decreased stigma, still so rampant in psychiatric illnesses, because patients see their problems as medical, not moral.
- SPECT increased patient compliance, as patients wanted to have better brains.

- SPECT led me to work with patients to slow down and in many cases reverse the brain aging process.

Over time, the imaging work led me to create a brain healthy business for our employees at the Amen Clinics and for other businesses. The collective health of your employees' brains are a business's most important asset. It also has led me to work on creating brain healthy churches, such as I am doing with the Daniel Plan at Saddleback Church. It has even helped to change the culture of our sports, as evidenced by our work with active and retired professional football players demonstrating very high levels of brain trauma.

Through our brain imaging work we have been able to see which factors hurt the brain and which ones help. Without directly looking at how the brain functions, we are just guessing at what is wrong with our patients and making way too many mistakes. How would you know about the health of your brain unless you looked?

WHEN SHOULD YOU THINK ABOUT GETTING A SPECT SCAN?

I think of SPECT like radar. Our Newport Beach clinic is very close to the Orange County / John Wayne Airport. On sunny days the pilot does not need radar to land the plane, because he can see the runway. But on days that are stormy, radar is an essential tool to safely land the plane. In the same way, when the clinical picture is clear a scan is not necessary. But for unclear or treatment-resistant cases, SPECT can provide useful, even lifesaving information.

Susan, forty-seven, came to see me for resistant depression. She had seen six other doctors and had tried ten different medications. She was suffering with severe depression, panic attacks, headaches, dizziness, and tremors. Everyone was treating her as a psychiatric case and changing her antidepressant medication and encouraging her to continue with psychotherapy that had no effect on her condition. She was feeling hopeless and suicidal, even though she had a large loving family, a supportive husband, and three teenage girls she adored.

Her SPECT scan looked awful. It showed overall severe decreased activity in a pattern that was consistent with an infection or toxic exposure.

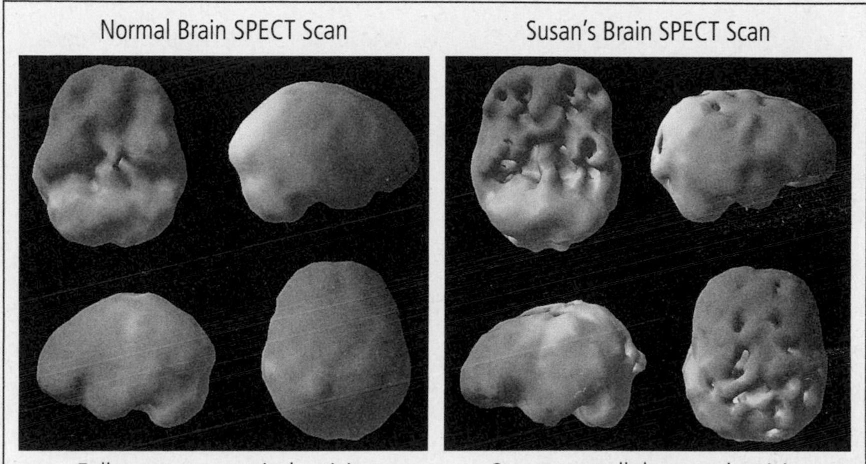

After an extensive workup we discovered that Susan had Lyme disease, for which she was able to get effective treatment. When I saw the scans it triggered more extensive questions that eventually led to the right diagnosis. Over time, with extensive brain rehabilitation, she did better. Susan later told me that the scan saved her life. Until she saw the damage in her brain she had just felt hopeless and helpless. Seeing her problem as medical and not moral made a big difference for her psychologically. It also pointed her medical team in the right direction. Too often when people do not get better, physicians diagnose them as "psychiatric" or with a personality problem. How would we know unless we looked?

Over the years it has become clear to me that the cost of having an ineffectively treated brain problem is much more expensive than a scan will ever be.

In our experience, more insurance companies are starting to pay for SPECT, especially for reasons such as memory problems, dementia, and traumatic brain injury. Initially insurance companies tend to deny new procedures, especially for mental health care. But the denial of SPECT for patients whose treatment may clearly benefit from it is a violation of the Paul Wellstone and Pete Domenici Mental Health Parity and Addiction Equity Act of 2008 (the Parity Act), which does not allow insurance carriers to discriminate against patients for mental health issues. They need to provide the same coverage as they do for medical procedures. Both the American College of Radiology and the European Society of Nuclear Medicine (ESNM) have published guidelines for using SPECT with a number of indications relevant to mental health issues, such as in cognitive decline, dementia, and traumatic brain injury. The ESNM guidelines also specifically reference SPECT's usefulness when evaluating psychiatric disorders. A recent analysis of 2,711 practice recommendations in cardiology, for example, found that only 11 percent were based on evidence from more than one randomized controlled clinical trial while 48 percent were based simply on expert opinion, case studies, or what was typically being done in practice. Similarly, a 2011 analysis behind the Infectious Diseases Society of America's practice guidelines found that 14 percent of the recommendations were based on a high level of scientific evidence. Medicine, in most places, is clearly still an art, not a hard science.

Applying a higher level of scientific evidence for SPECT for mental health issues than to other covered medical procedures is a violation of the Parity Act. The Parity Act states: "The treatment limitations applicable to such mental health or substance use disorder benefits are no more restrictive than the predominant treatment limitations applied to substantially all medical and surgical benefits covered by the plan (or coverage) and there are no separate treatment limitations that are applicable only with respect to mental health or substance use disorder benefits."

WHAT CAN I DO I IF I CANNOT GET A SCAN?

I have known for a long time that many people cannot get scans, either because of the cost or because they are not near a center that performs them. My books are translated into thirty languages, and if you read one in China or Brazil, odds are you are not going to get a scan. So based on thousands of scans, we developed a series of questionnaires to help people predict what their scan *might* look like if they could get one. Then, based on the answers, we give suggestions on ways to think about helping the brain with certain natural supplements, medications, or exercises. For uncomplicated cases, this questionnaire has proven to be surprisingly accurate, usually lining up with what we see on scan very well. The questionnaires can be found online at www.amenclinics.com or www.theamensolution.com and they are used by mental health professionals around the world. Of course, you should always talk with your own health care professional when considering treatment options.

BE A BRAIN WARRIOR FOR YOURSELF AND THOSE YOU LOVE

Brain SPECT imaging has also taught me that I have to be a "brain warrior" for my own health. No one is going to do it for me. In fact, others will try to steal your brain health in search of their monetary gain: "Do you want to supersize your fries for only a few more pennies?" You would do well to be a warrior for your own brain health.

- Be a brain warrior by boosting your brain to dramatically increase the quality and consistency of your decisions.
- Be a brain warrior by always working to be more conscientious and thoughtful regarding your health.

- Be a brain warrior by protecting your brain from injuries and toxins.
- Be a brain warrior by saying no to the food pushers in your life who ask you to supersize, refill, have extra helpings, or eat more than you need to keep your body and brain healthy.
- Be a brain warrior by getting and keeping your weight at a healthy level.
- Be a brain warrior by only eating food that serves you rather than making food companies more profitable.
- Be a brain warrior by getting the sleep you need.
- Be a brain warrior by increasing your endurance and strength through smart exercise.
- Be a brain warrior by embracing constant mental exercise and new learning.
- Be a brain warrior by getting problems like ADHD, anxiety, depression, and other mental health problems treated. Start with natural treatments first if it makes sense.
- Be a brain warrior by killing the ANTs (automatic negative thoughts) that steal your happiness and make you old.
- Be a brain warrior by developing a consistent stress reduction practice.
- Be a brain warrior by rehabilitating your brain if it has been hurt or you have been bad to it up until now.
- Be a brain warrior by taking your simple supplements to give it the nutrients it needs.
- Be a brain warrior by building your own genius network of supporters. The health of the people you spend time with matters to your health.
- Be a brain warrior by knowing your important numbers, such as blood pressure, vitamin D level, and HgA1C, to make sure they are in the healthy range.

- Be a brain warrior by giving the gift of brain health to your children, grandchildren, family, and friends.
- Be a brain warrior by getting your inner child under control!

CHANGE YOUR AGE NOW: TWENTY BRAIN TIPS TO IMPROVE YOUR LIFE FROM BRAIN SPECT RESEARCH

1. Brain envy often results from seeing your own brain SPECT scan. When I saw my scan for the first time in 1991, I realized that even though I never used any drugs, rarely drank alcohol, and never smoked, my brain needed help. I wanted a healthier brain. Determine to do whatever you can to give yourself the best brain possible, because when your brain works right, you work right.

2. Brains usually become less active with age. But they do not have to. With a brain-smart program, you can start to reverse the brain aging process.

3. When I exchanged my bad brain habits for good ones, my brain actually aged backward. I have seen evidence of this on my SPECT scans over the years: It looks functionally younger today than it did fifteen years ago. I've seen this happen to scores of our own clients. You too can decrease your brain's functional age as you determine to trade bad brain habits for good ones.

4. Many "unexpected" things can hurt the brain, such as being overweight, eating lots of sugar, using household chemicals without proper ventilation, getting less than seven hours of sleep, overworking, being around secondhand smoke, and drinking diet sodas. Small changes over time can add up to big improvements in brain function.

5. Before I became aware of brain health, I thought that some of my cognitive issues, such as dropping names or being unfocused and forgetful, were just a normal part of the aging process. At thirty-seven! Dropping names or being unfocused or forgetful at any age is a sign of trouble.

6. Having a less-than-healthy brain goes with less-than-healthy decisions and a suboptimized life.

7. You cannot change what you do not measure.

8. Seeing a SPECT scan of your own brain or of someone you love increases understanding and compassion. If someone has a brain that is misfiring, they are literally working with an emotional and cognitive handicap. They deserve help and compassion, not condemnation. If the brain is balanced, better behavior tends to follow.

9. SPECT scans do not lie. If someone is in denial about the damage they are doing to their brain by drinking too much or using drugs or trying to hide their addiction, a scan will reveal truth. Getting a brain scan often serves as the impetus for going to rehab or AA, or simply making a once-and-for-all decision to quit harming their brain.

10. If you want to live a long time with your mind intact, it is critical to become a brain warrior for your own health. No one is going to do it for you. You have to take custody of your own brain health. Begin by building your own genius network of supporters. The health of the people you spend time with matters to your health.

11. Be a brain warrior by getting and keeping your weight at a healthy level. Eat food that serves you, exercise regularly, and give your brain simple supplements to nourish it well.

12. Be a brain warrior by getting the sleep you need, dealing with sleep apnea, and taking sleep-friendly supplements like melatonin and GABA for a good night's sleep, and using sleep-inducing routines to prepare for bed.

13. Be a brain warrior by developing a consistent stress reduction practice with meditation or deep breathing or contemplative prayer. Kill the ANTs that steal your happiness and make you old.

14. Be a brain warrior by protecting your brain from injuries and toxins. Start a brain healthy rehabilitation program if you've damaged your brain or suffered any kind of brain injury.

15. Be a brain warrior by constant mental exercise and new learning. Keep your childlike curiosity and openness to new knowledge intact to stay young of mind and heart.

16. Be a brain warrior by being an example of good physical and mental health for your family and friends. Pass on the encouragement!

17. A true sign of self-love is how well you take care of your brain and body. If you are not doing it, why not? You are worth it!

18. Success breeds success. When you do the right things over time they become easier and easier to do.

19. The quality of your life is determined by the sum of all of the decisions in your life. With a better brain you are much more likely to make better decisions and dramatically improve everything in your life.

20. Even if you have been bad to your brain, you can literally improve your brain, and when you do you improve everything in your life.

APPENDIX

NATURAL SUPPLEMENTS

To Enhance Your Brain and Extend Your Life

Using supplements by themselves, without getting your diet, exercise, thoughts, peer group, and environment under control, is a waste of money. You need to do the whole program for it to work. Yet I have seen instances in which supplements used in conjunction with a brain-smart plan make a significant difference.

Let me start by explaining the pros and cons of using natural supplements to enhance brain function. To start, they are often effective. They usually have dramatically fewer side effects than most prescription medications and they are significantly less expensive. Plus, you never have to tell an insurance company that you have taken them. As awful as it sounds, taking prescription medications can affect your insurability. I know many people who have been denied or made to pay higher rates for insurance because they have taken certain medications. If there are natural alternatives, they are worth considering.

Yet natural supplements also have their own set of problems. Even though they tend to be less expensive than medications, they may be more expensive for you personally because they are usually not covered by insurance. Many people are unaware that natural supplements can have side effects and need to be thoughtfully used. Just because something is "natural" does not mean it is innocuous. Both arsenic and cyanide are natural, but that doesn't mean they are good for you. For example, St. John's Wort, one of my favorite natural antidepressants, can cause sun sensitivity, and it can also decrease the effectiveness of a number of medications such as birth control pills. Oh, great! Get depressed, take St. John's Wort from the grocery store, and now you are pregnant when you don't want to be. That may not be a good thing.

One of the major concerns about natural supplements is the lack of quality control. There is variability among brands, so you need to find brands you trust. Another disadvantage is that many people get their advice about supplements from the teenage clerk at the health food store, who may not have the best information. But, even when looking at the problems, the benefits of natural supplements make them worth considering, especially if you can get thoughtful, research-based information.

Every day I personally take a handful of supplements, which I believe make a significant difference in my life. They have helped to change the health of my brain, my energy, and my lab values. Many physicians say that if you eat a balanced diet you do not need supplements. I love what Dr. Mark Hyman wrote in his book *The UltraMind Solution: Fix Your Broken Brain by Healing Your Body First:* If people "eat wild, fresh, organic, local, non–genetically modified food grown in virgin mineral- and nutrient-rich soils that has not been transported across vast distances and stored for months before being eaten . . . and work and live outside, breathe only fresh unpolluted air, drink only pure, clean water, sleep nine hours a night, move their bodies every day, and are free from chronic stressors and exposure to environmental toxins," then it is possible that they might not need supplements. Because we live in a fast-paced society where we pick up food on the fly, skip meals, eat sugar-laden treats, buy processed foods, and eat foods that have been chemically treated, we could all use a little help from a multivitamin/mineral supplement.

AMEN SOLUTION SUPPLEMENTS

At the Amen Clinics we make our own line of supplements, the Amen Solution, that have taken over a decade to develop. The reason I developed this line was that I wanted my patients and my own family to have access to the highest-quality research-based supplements available. After I started recommending supplements to my patients, they would go to the supermarket, drugstore, or health food store and face so many choices that they did not know what or how to choose. This dilemma was compounded by the varying levels of quality among different brands.

Another reason I developed my own line was that the Amen Clinics see a high number of people who have ADD. I realized if they did not get their supplements as they walked out the door, they would forget about it or procrastinate and not have started them by their next appointment.

Research shows the therapeutic benefit of using supplements to support a healthy mood, sleep, and memory. I strongly recommend that when purchasing a supplement, you consult a health care practitioner familiar with nutritional supplements to determine which supplements and dosages may be most effective for you. Our website (www.amenclinics.com) contains links to the scientific literature on many different supplements related to brain health, so that you, as a consumer, can be fully informed on the benefits and risks involved. Please remember supplements can have very powerful effects on the body, and caution should be used when combining them with prescription medications.

THREE SUPPLEMENTS FOR EVERYBODY

There are three supplements I typically recommend to *all* of my patients because they are critical to optimal brain function: a multivitamin, fish oil, and vitamin D.

Multivitamins. According to recent studies, more than 50 percent of Americans do not eat at least five servings of fruits and vegetables a day, the minimum required to get the nutrition you need. I recommend that all of my patients take a high-quality multivitamin/mineral complex every day. In an editorial in the *Journal of the American Medical Association,* researchers recommended a daily vitamin for everybody because it helps prevent chronic illness. In addition, people with weight-management issues often are not eating healthy diets and have vitamin and nutrient deficiencies. Furthermore, research suggests that people who take a multiple vitamin actually have younger-looking DNA.

A 2010 study from Northumbria University in England tested multivitamins' effects on 215 men between the ages of thirty and fifty-five. For the double-blind, placebo-controlled study, the men were tested on mental performance and asked to rate themselves on general mental health, stress, and mood. At the debut of the trial, there were no significant differences between the multivitamin group and the placebo group. When the participants were retested a little more than one month later, the multivitamin group reported improved moods and showed better mental performance, helping participants be happier and smarter! Not only that, the multivitamin group reported an improved sense of vigor, reduced stress, and less mental fatigue after completing mental tasks.

Another placebo-controlled study from Northumbria researchers tested the effects of multivitamins on eighty-one healthy children aged eight to fourteen. They found that the children who took multivitamins performed better on two out of three attention tasks. The researchers concluded that multivitamins have the potential to improve brain function in healthy children.

NeuroVite Plus is the brand we make at the Amen Clinics. It contains a complete range of brain healthy nutrients. Four capsules a day is the full dose, which contains the following:

- Vitamin A and high levels of Bs, plus vitamins C, D (2,000 IUs), E, and K_2
- Minerals, including zinc, copper, magnesium, selenium, chromium, manganese, calcium and magnesium

- Brain nutrients—alpha-lipoic acid, acetyl-L-carnitine, and phosphatidylserine
- Equivalent nutrients to:
  - 1 apple (quercitin)
  - 1 tomato (lycopene)
  - 1 serving fresh spinach (lutein)
  - 1 serving broccoli (broccoli seed concentrate)
  - 2 L red wine (resveratrol, without the alcohol)
  - a cup of blueberries (pterostilbene)
- 1 full dose stabilized probiotic

Fish oil. For years, I have been writing about the benefits of omega-3 fatty acids, which are found in fish oil supplements. I personally take a fish oil supplement every day and recommend that *all* of my patients do the same. When you look at the mountain of scientific evidence, it is easy to understand why. Research has found that omega-3 fatty acids are essential for optimal brain and body health.

For example, according to researchers at the Harvard School of Public Health, having low levels of omega-3 fatty acids is one of the leading preventable causes of death and has been associated with heart disease, strokes, depression, suicidal behavior, ADD, dementia, and obesity. There is also scientific evidence that low levels of omega-3 fatty acids play a role in substance abuse.

I can tell you that most people, unless they are focusing on eating fish or taking fish oil supplements, have low omega-3 levels. I know this because at the Amen Clinics we perform a blood test on patients where we can measure the levels of omega-3 fatty acids in the blood. Before I began offering the test to patients, I tested it on my employees, several family members, and of course, myself. When my test results came back, I was very happy with the robust numbers. An omega-3 score above 7 is good. Mine was nearly 11. But the results for nearly all of the employees and family members I tested were not so good. In fact, I was horrified at how low their levels were, which put them at greater risk for both physical and emotional problems. It is an easy fix. They just needed to eat more fish or take fish oil supplements.

Boosting your intake of omega-3 fatty acids is one of the best things you can do for your brainpower, mood, and weight. The two most studied omega-3 fatty acids are eicosapentaenoic acid (EPA) and docosahexaenoic acid (DHA). DHA makes up a large portion of the gray matter of the brain. The fat in your brain forms cell membranes and plays a vital role in how our cells function. Neurons are also rich in omega-3 fatty acids. EPA improves blood flow, which boosts overall brain function.

Increasing omega-3 intake has been found to decrease appetite and cravings and reduce body fat. In a fascinating 2009 study in the *British Journal of Nutrition,* Australian researchers analyzed blood samples from 124 adults (21 healthy weight, 40 overweight, and 63 obese), calculated their BMI, and measured their waist and hip circumference. They found that obese individuals had significantly lower levels of EPA and DHA compared with healthy-weight people. Subjects with higher levels were more likely to have a healthy BMI and waist and hip measurements.

More evidence about the benefits of fish oil on weight loss comes from a 2007 study from the University of South Australia. The research team found that taking fish oil combined with moderate exercise, like walking for forty-five minutes three times a week, leads to a significant reduction in body fat after just twelve weeks. But taking fish oil without exercising, or exercising without fish oil, did not result in any reduction in body fat.

One of the most intriguing studies I have found on fish oil and weight loss appeared in a 2007 issue of the *International Journal of Obesity.* In this study, researchers from Iceland investigated the effects of seafood and fish oils on weight loss in 324 young overweight adults with BMIs ranging from 27.5 to 32.5. The participants were placed in four groups that ate 1,600-calorie diets that were the same except that each group's diet included only one of the following:

- Control group (sunflower oil capsules, no seafood or fish oil)
- Lean fish (3 × 150 g portions of cod per week)
- Fatty fish (3 × 150 g salmon per week)
- Fish oil (DHA/EPA capsules, no seafood)

After four weeks, the average amount of weight loss among the men in each of the four groups was as follows:

- Control group: 7.8 pounds
- Lean fish group: 9.6 pounds
- Fatty fish group: 9.9 pounds
- Fish oil group: 10.9 pounds

The researchers concluded that adding fish or fish oil to a nutritionally balanced calorie-restricted diet could boost weight loss in men.

Research in the last few years has also revealed that diets rich in omega-3 fatty acids help promote a healthy emotional balance and positive mood in later years, possibly because DHA is a main component of the brain's synapses. A growing body of scientific evidence indicates that fish oil helps ease symptoms of depression. One twenty-year study involving 3,317 men and women found that people with the highest consumption of EPA and DHA were less likely to have symptoms of depression.

There is a tremendous amount of scientific evidence pointing to a connection between the consumption of fish that is rich in omega-3 fatty acids and cognitive function. A Danish team of researchers compared the diets of 5,386 healthy older individuals and found that the more fish in a person's diet, the longer the person was able to maintain their memory and reduce the risk of dementia. Dr. J. A. Conquer and colleagues from the University of Guelph in Ontario studied the blood fatty acid content in the early and later stages of dementia and noted low levels when compared with healthy people. In 2010, UCLA researchers analyzed the existing scientific literature on DHA and fish oil and concluded that supplementation with DHA slows the progression of Alzheimer's and may prevent age-related dementia.

Omega-3 fatty acids benefit cognitive performance at every age. Scientists at the University of Pittsburgh reported in 2010 that middle-aged people with higher DHA levels performed better on a variety of tests, including nonverbal reasoning, mental flexibility, working memory, and vocabulary. A team of Swedish researchers surveyed nearly five thousand fifteen-year-old boys and found that those who ate fish more than once a week scored higher on standard intelligence tests than teens who ate no fish. A follow-up study found that teens eating fish more than once a week also had better grades at school than students with lower fish consumption.

Additional benefits of omega-3 fatty acids include increased attention in people with ADD, reduced stress, and a lower risk for psychosis. When we put our retired football players on our fish oil supplements, many of them were able to decrease or completely eliminate their pain medications.

My recommendation for most adults is to take 1–2 g high-quality fish oil a day balanced between EPA and DHA.

Omega 3 Power is our brand to support healthy brain and heart function by providing highly purified omega-3 fatty acids (EPA and DHA) from the most advanced production, detoxification, and purification process in the industry. It is produced under the natural industry's most rigorous standards. Each batch of our oil is independently analyzed by the third-party lab Eurofins for more than 250 environmental contaminants, including PCBs. Our oil is certified to be more than twenty times lower than California's Prop 65 requirement of fewer than 90 nanograms/day, and it exceeds all other domestic and international regulatory standards. Two softgel tablets contain 2.8 g fish oil, 860 mg EPA, and 580 mg DHA.

Vitamin D. Vitamin D, sometimes called the sunshine vitamin, is best known for building bones and boosting the immune system. But it is also an essential vitamin for brain health, mood, memory, and your weight. While classified as a vitamin, it is a steroid hormone vital to health. Low levels of vitamin D have been associated with depression, autism, psychosis, Alzheimer's disease, MS, heart disease, diabetes, cancer, and obesity. Unfortunately, vitamin D deficiencies are becoming more and more common, in part because we are spending more time indoors and using more sunscreen.

Did you know that when you don't have enough vitamin D, you feel hungry all the time, no matter how much you eat? That is because low levels of vitamin D interfere with the effectiveness of leptin, the appetite hormone that tells you when you are full. Research also shows that vitamin D insufficiency is associated with increased body fat. A 2009 study out of Canada found that weight and body fat were significantly lower in women with normal vitamin D levels than women with insufficient levels. It appears that extra fat inhibits the absorption of vitamin D. The evidence shows that obese people need higher doses of vitamin D than lean people to achieve the same levels.

One of the most interesting studies I have seen on vitamin D comes from researchers at Stanford Hospital and Clinics. They detailed how a patient was given a prescription for 50,000 IU/week vitamin D that was incorrectly filled for 50,000 IU/*day.* After six months, the patient's vitamin D level increased from 7, which is extremely low, to 100, which is at the high end of normal.

What I found really intriguing about this report was that the patient complained of a few side effects from the very high dosage, namely decreased appetite and significant weight loss. Of course, I am not advocating that you take more vitamin D than you need. But I think it shows that optimal levels of vitamin D may play a role in appetite control and weight loss.

This patient's story shows why it is so important to get your vitamin D level checked before and after treatment. That way, you will know if you are taking the right dosage, or if you need to adjust it.

Vitamin D is so important to brain function that its receptors can be found throughout the brain. Vitamin D plays a critical role in many of the most basic cognitive functions, including learning and making memories. These are just some of the areas where vitamin D affects how well your brain works, according to a 2008 review that appeared in the *FASEB Journal.*

The scientific community is waking up to the importance of vitamin D for optimal brain function. In the past few years, I have come across a number of studies linking a shortage of vitamin D with cognitive impairment in older men and women, as well as some suggesting that having optimal levels of the sunshine vitamin may play a role in protecting cognitive function. One such study in the *Journal of Alzheimer's Disease* found that vitamin D_3, the active form of vitamin D, may stimulate the immune system to rid the brain of beta-amyloid, an abnormal protein that is believed to be a major cause of Alzheimer's disease. Vitamin D activates receptors on neurons in regions important in the regulation of behavior, and it protects the brain by acting in an antioxidant and anti-inflammatory capacity.

Another study conducted in 2009 by a team at Tufts University in Boston looked at vitamin D level in more than a thousand elderly people over the age of sixty-five and its effect on cognitive function. Only 35 percent of the participants had optimal vitamin D levels; the rest fell in the insufficient or deficient categories. The individuals with optimal levels of vitamin D (50 nmol/l or higher) performed better on tests of executive functions, such as reasoning, flexibility, and perceptual complexity. They also scored higher on attention and processing speed tests than their counterparts with suboptimal levels.

The lower your vitamin D levels, the more likely you are to feel blue rather than happy. Low levels of vitamin D have long been associated with a higher incidence of depression. In recent years, researchers have been asking if, given this association, vitamin D supplementation can improve moods.

One trial that attempted to answer that question followed 441 overweight and obese adults with similar levels of depression for one year. The individuals took either a placebo or one of two doses of vitamin D: 20,000 IU per week or 40,000 IU per week. By year's end, the two groups that had taken the vitamin D showed a significant reduction in symptoms while the group taking the placebo reported no improvements. Other trials have reported similar findings.

The current recommended dose for vitamin D is 400 IU daily, but most experts agree that this is well below the physiological needs of most individuals and instead suggest 2,000 IU of vitamin D daily. I think it is very important to test your individual needs, especially if you are overweight or obese, since your body may not absorb the vitamin D as efficiently if you are heavier.

Vitamin D_3, comes as 1,000 IU tablets, 2,000 IU tablets, or as 10,000 IU liquid.

BRAIN AND MEMORY POWER BOOST

This is the supplement formulated to help in our brain enhancement work with active and retired NFL players. When used in conjunction with a brain healthy program, we demonstrated significant improvement in memory, reasoning, attention, processing speed, and accuracy. It was so effective that I take it every day.

Brain and Memory Power Boost includes the super-antioxidant

N-acetylcysteine (NAC), along with phosphatidylserine to maintain the integrity of cell membranes, and huperzine A and acetyl-L-carnitine to enhance acetylcholine availability, and vinpocetine and ginkgo biloba to enhance blood flow. This is a novel combination of powerful antioxidants and nutrients essential in enhancing and protecting brain health. It supports overall brain health, circulation, memory, and concentration.

CRAVING CONTROL

The key to successful weight management is eating a brain healthy diet and managing your cravings. In support of this goal, Craving Control was developed, a powerful new nutritional supplement formulated to support healthy blood sugar and insulin levels while providing antioxidants and nutrients to the body. Our formulation includes NAC and glutamine to reduce cravings, chromium and alpha-lipoic acid to support stable blood sugar levels, and a brain healthy chocolate and D,L-phenylalanine designed to boost endorphins.

This is the formula we use at the Amen Clinics in our own weight-loss groups. In the first group, participants who used the craving formula and attended each group lost an average of 10 pounds in ten weeks.

RESTFUL SLEEP

Sleep is essential to healthy brain function. Restful Sleep is formulated with a combination of nutrients designed to support a calm mind and promote a deep, relaxed, restful night's sleep. This supplement contains both immediate and time-released melatonin to keep you asleep throughout the entire night, plus the calming neurotransmitter GABA, a combination of the essential elements zinc and magnesium, and the herb valerian, which together may produce an overall sedative effect to help support sleep. At the Amen Clinics we refer to Restful Sleep as "the hammer" because so many people have told us it has helped them.

SAMe MOOD AND MOVEMENT SUPPORT

SAMe has scientific research suggesting it helps support mood, movement, and pain control. It is intimately involved in the creation of the key neurotransmitters, serotonin, dopamine and norepinephrine, which support healthy mood. And as an added advantage, SAMe has also been shown to support healthy joints and decrease pain. The typical dose is anywhere from 400 mg to 800 mg twice a day. It is usually better to take earlier in the day, as it can be energizing. Research suggests you should be cautious with SAMe if you have bipolar disorder.

SEROTONIN MOOD SUPPORT

Serotonin Mood Support promotes normal serotonin levels by providing 5-HTP, a direct precursor to serotonin, along with a proprietary extract of saffron, shown clinically to support a normal mood. Vitamin B_6 and inositol are included to provide additional synergistic support. Serotonin Mood Support is useful to support a healthy mood when serotonin levels are suspected to be low. It seems to be especially helpful for people who tend to get stuck on negative thoughts or negative behaviors. It has also been shown to help support healthy sleep patterns.

FOCUS AND ENERGY OPTIMIZER

Formulated without caffeine that makes people jittery, Focus and Energy Optimizer supports both focus and healthy energy levels. It is formulated with green tea and choline to help with focus, along with three powerful adaptogens, which act synergistically to enhance endurance and stamina. The adaptogens ashwagandha, rhodiola, and panax ginseng have been scientifically shown to improve the body's resistance to stress and support a healthy immune system.

GABA CALMING SUPPORT

GABA Calming Support promotes natural relaxation and calm by providing a combination of inhibitory neurotransmitters vital to quieting an overactive mind. It contains clinically tested and natural Pharma GABA shown to promote relaxation by increasing calming, focused brain waves while also reducing other brain waves associated with worry. Complementing this clinically tested and natural substance are vitamin B_6, magnesium, and lemon balm, an herb traditionally known for its calming effects.

ROBERT

Robert was a defensive back for the Minnesota Vikings. He is tall, lean, and in seemingly good health. When he joined our professional football brain trauma/rehabilitation study he complained that his memory was not as good as it was before and he was having to keep more notes. His main concern was that he heard that many former NFL players were struggling with memory problems, far more than others, and he had a parent who had been diagnosed with Alzheimer's disease, increasing his own vulnerability and anxiety.

Robert's initial SPECT scan showed highly significant levels of brain damage, especially to his prefrontal cortex (judgment), temporal lobes (memory), parietal lobes (direction sense), and cerebellum (coordination). On our memory test, his memory scored in the 5th percentile, which means that 95 percent of people his age and education scored better than Robert, who was a Stanford graduate.

To our delight, Robert's follow-up brain SPECT scan was dramatically better and his memory test improved 1,000 percent. How? One of the best things about Robert is that he was compliant with all of our directions. He faithfully took our multivitamin, NeuroVite; fish oil; Omega-3 Power; and our brain support supplement, Brain and Memory Power Boost. He did not miss doses and was consistent throughout the period of the study. His follow-up testing showed that his memory improved 1,000 percent and he scored in the 55th percentile compared with his peers. His SPECT scan showed dramatic improvement in all of the areas that were problematic.

Robert's Before Brain SPECT Scan | Robert's After Brain SPECT Scan

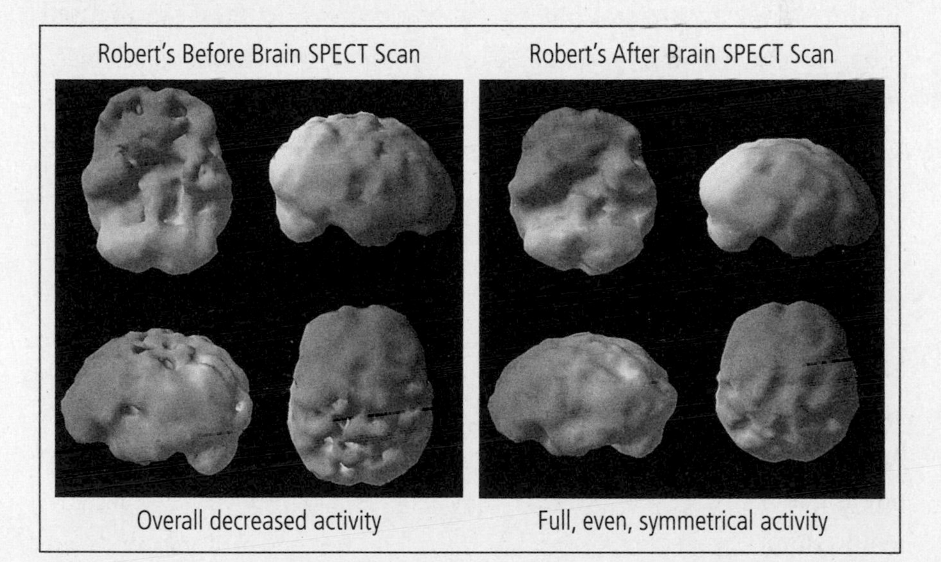

Overall decreased activity | Full, even, symmetrical activity

Before his first evaluation Robert was clearly headed for trouble. With these simple interventions, his brain and test scores have shown remarkable improvement. His brain has literally aged backward. I am very excited for Robert's progress. He took thousands of hits to his head playing high school, college, and professional football, some thirty years before we first saw him. Yet, despite the damage and the distance in years, his brain still showed a remarkable ability to recover.

The good news about our study is that it has demonstrated the damaged brain's ability to show high levels of improvement on a simple, inexpensive, smart program.

NOTE ON REFERENCES AND FURTHER READING

The information in *Use Your Brain to Change Your Age* is based on more than four hundred sources, including scientific studies, books, interviews with medical experts, statistics from government agencies and health organizations, and other reliable resources. Printed out, the references take up more than sixty pages. In an effort to save a few trees, I have decided to place them exclusively on the Amen Clinics website. I invite you to view them at www.amenclinics.com/uybcya.

ACKNOWLEDGMENTS

I live in gratitude for having an amazing group of colleagues and friends who have helped me with this work. I am especially grateful to all of my patients and friends who have allowed me to share their stories with you. Thank you to Dr. Doris Rapp, Steve, Marianne, and Carlos for allowing me to tell your inspiring stories.

Thank you to Pastor Rick Warren who trusted me, Dr. Mehmet Oz, and Dr. Mark Hyman for the opportunity to help design the Daniel Plan for Saddleback Church. I may be the only person to congratulate a pastor on helping his church's numbers to shrink (on the scales, that is)—but a 250,000-pound weight loss is a breakthrough worth being excited about—especially since the weight dropped off via a high-nutrition eating plan, which left participants not only thinner but healthier and more energetic too. Thank you to Pastor Warren's assistant, Steve Komanapalli, for opening your big heart as you so honestly shared your story of changing your health habits, so you can leave a healthier legacy for your family. Dee Eastman and Debbie Eaton: Thank you both for letting us in on what you've learned about the power of small groups to change lives at Saddleback and beyond. You ladies do an incredible job. Joe Polish: Thank you, my friend, for creating the "Create Your Own Genius Network" exercise and allowing me to pass it with our readers.

Kudos to my sister-in-law, Tamara, for sharing her inspiring story of transformation, and to her sister, Tana (who happens to be my lovely wife), for her compassion and encouragement in the process. Dr. Riz Malik is a gifted psychiatrist who works at our Amen Clinics in Reston, Virginia. Riz, you made my day with the e-mail you sent titled "A

Different Person." Thank you for describing your journey to health and fitness with us and now to the readers of this book.

Dr. Andy McGill, I am so grateful to you for sharing how you managed to turn your life around in midlife, thus allowing you to have a younger brain today than you had ten years ago. I sense many will do a 180-degree turnaround for the better after reading your story. Dr. Joe Dispenza, have I told you lately how brilliant you are? The insights you gave us on how to make a once-and-for-all decision to change are simply amazing and so appreciated. Dr. Cyrus Raji, I believe your magnificent research on exercise and Alzheimer's will motivate thousands to get walking in order to shrink their bodies and grow their brains. So grateful for the good work you do.

Jim Kwik, your contribution to lifelong learning and to this book is incredible. Thank you for being so generous with your heart and mind. Thank you to Savannah DeVarney, who shared her insights on our 24/7 Brain Gym at www.amensolutions.com, and great information about the "why" of regularly exercising the mind.

Joni Houtain, everyone should have a cheerleader like you in their lives. Thank you for sharing your story with such vulnerability and good humor. Laughter is good medicine. May you have many more years of "growing young" in body and soul.

Chris Hartsfield, you have suffered the worst a parent can imagine, and now you are helping others who are struggling to find a way to thrive after great sorrows. Your story of Sammie's journey is a treasure, shining light in dark places, and showing that we can find joy and health, even after unspeakable loss. Your story will change lives, perhaps even save some. Gerald Sharon, blessings to you for encouraging others to tend to their health while they are going through stress and grief. You honor your wife's memory by taking such good care of yourself.

Gratitude to all the football players and legends who participated in our NFL study, with a special shout-out to AD (Anthony Davis), Roy Williams, Marvin Fleming, Fred Dryer, and Cam Cleeland for letting us see their stories up close and personal. Also, I send my gratitude to Captain Patrick Caffrey for sharing his story and passion to help wounded warriors, and to Ray and Nancy, who continually inspire me to keep doing what I do.

I am especially grateful to Becky Johnson and Frances Sharpe, who were invaluable in the process of researching, interviewing, and completing this book. Also, our research department, including Dr. Kristen Willeumeir and Derek Taylor, provided valuable insights and encouragement. Other staff at Amen Clinics, Inc., as always, provided tremendous help and support during this process, especially my personal assistant, Catherine Hanlon, and Dr. Joseph Annibali. I am also grateful to my friend and colleague Dr. Earl Henslin, who read the manuscript and gave thoughtful suggestions.

I also wish to thank my amazing literary team at Crown Archetype, especially my kind and thoughtful editor, Julia Pastore, and my publisher, Tina Constable. I am forever grateful to my literary agent, Faith Hamlin, who besides being one of my best friends, is a thoughtful, protective, creative mentor, along with Stephanie Diaz, our foreign rights agent. If you are reading this outside of the United States, Stephanie made that happen. In addition, I am grateful to all of my friends and colleagues at public television stations across the country. Public television is a treasure to our country and I am grateful to be able to partner with stations to bring our message of hope and healing to you. And, to Tana—my wife, my joy, and my best friend—who patiently listened to me for hours on end and gave many thoughtful suggestions on the book. I love all of you.

INDEX

ABOUT DANIEL G. AMEN, M.D.

Dr. Amen is a physician, psychiatrist, teacher, and four-time *New York Times* bestselling author. He is widely regarded as one of the world's foremost experts on applying brain imaging science to clinical psychiatric practice. He is a board-certified child and adult psychiatrist and a Distinguished Fellow of the American Psychiatric Association. He is the medical director of Amen Clinics, Inc., in Newport Beach and San Francisco, California; Bellevue, Washington; and Reston, Virginia. Amen Clinics has the world's largest database of functional brain scans related to behavior, totaling more than seventy thousand scans. The clinics have seen patients from ninety countries.

Dr. Amen is an assistant clinical professor of psychiatry and human behavior at the University of California Irvine School of Medicine. He is widely regarded as a gifted teacher, taking complex concepts in neuropsychiatry, neuroimaging, and brain health to make them easily accessible to other professionals and the general public.

Dr. Amen is the lead researcher on the world's largest brain imaging/brain rehabilitation study on professional football players, which not only demonstrated significant brain damage in a high percentage of retired players but also the possibility for rehabilitation in many with the principles that underlie his work.

Under the direction of Pastor Rick Warren, Dr. Amen together with Drs. Mark Hyman and Mehmet Oz, is one of the chief architects of Saddleback Church's Daniel Plan, a fifty-two-week program to get churches healthy, physically, emotionally, and spiritually.

Dr. Amen is the author of forty-five professional articles; the co-author of the chapter "Functional Imaging in Clinical Practice" in the *Comprehensive Textbook of Psychiatry;* and the author of twenty-eight books, including *Change Your Brain, Change Your Life; Magnificent Mind at Any Age; Change Your Brain, Change Your Body;* and *The Amen Solution.* He is also the author of *Healing ADD, Making a Good Brain Great,* and *Healing the Hardware of the Soul,* and coauthor of *Unchain Your Brain, Healing Anxiety and Depression,* and *Preventing Alzheimer's.*

Dr. Amen is the producer and star of six highly popular shows about the brain, which have raised more than thirty-five million dollars for public television.

A small sample of the organizations Dr. Amen has spoken for includes the National Security Agency; the National Science Foundation; Harvard's Learning and the Brain Conference; Franklin Covey; the Million Dollar Roundtable; the National Council of Juvenile and Family Court Judges; and the Supreme Courts of Delaware, Ohio, and Wyoming. Dr. Amen's work has been featured in *Newsweek, Parade* magazine, the *New York Times Magazine, Men's Health,* and *Cosmopolitan.*

Dr. Amen is married to Tana; he is the father of four children and grandfather to Elias, Julian, Angelina, Emmy, and Liam. He is an avid table tennis player.

ABOUT AMEN CLINICS, INC.

Amen Clinics, Inc. (ACI) was established in 1989 by Daniel G. Amen, M.D. It specializes in innovative diagnosis and treatment planning for a wide variety of behavioral, learning, emotional, cognitive, and weight problems for children, teenagers, and adults. ACI has an international reputation for evaluating brain-behavior problems, such as ADD, depression, anxiety, school failure, brain trauma, obsessive-compulsive disorders, aggressiveness, marital conflict, cognitive decline, brain toxicity from drugs or alcohol, and obesity.

Brain SPECT imaging is performed in the Clinics. ACI has the world's largest database of brain scans for emotional, cognitive, and behavioral problems. ACI welcomes referrals from physicians, psychologists, social workers, marriage and family therapists, drug and alcohol counselors, and individual clients.

Amen Clinics, Inc., Newport Beach
4019 Westerly Pl., Suite 100
Newport Beach, CA 92660
(888) 564–2700

Amen Clinics, Inc., San Francisco
1000 Marina Blvd., Suite 100
Brisbane, CA 94005
(888) 564–2700

Amen Clinics, Inc., Northwest
616 120th Ave. NE, Suite C100
Bellevue, WA 98005
(888) 564–2700

Amen Clinics, Inc., DC
1875 Campus Commons Dr.
Reston, VA 20191
(888) 564–2700
www.amenclinics.com

WWW.AMENCLINICS.COM

Amenclinics.com is an educational, interactive website geared toward mental health and medical professionals, educators, students, and the general public. It contains a wealth of information and resources to help you learn about and optimize your brain. The site contains more than three hundred color brain SPECT images, thousands of scientific abstracts on brain SPECT imaging for psychiatry, a free brain healthy audit, and much more.